2\95 P

D0984783

Problem Solving and Flowcharting

Problem Solving and Flowcharting

Ronald E. Elliott
College of the Mainland
Texas City, Texas

RESTON PUBLISHING COMPANY, INC., *Reston, Virginia*

10 9 8 7 6 5 4 3 2 1

ISBN: 0-87909-643-8 (P)
0-87909-644-6 (C)

Library of Congress Catalog Number: 74-183131
Printed in the United States of America

Preface

Flowcharting is the first and most basic step in problem solving. This applies to problems in general and particularly to computer applications. This book is intended to help you learn the logic involved in drawing flowcharts. The flowcharting is not an end in itself, but rather a means to the end of finding a solution. Therefore, the primary emphasis is on the logic involved, while emphasis on technique is held to a minimum. Once the logic is mastered, the technique will follow.

Problems are presented at the end of each chapter (except the first) partly to apply the principles brought forth in that chapter, and partly to help develop your problem solving ability. Most of the problems are "word" problems because most problems occur as words rather than equations. Some of the problem statements, particularly in the latter chapters, are somewhat ill-defined. This is because, again, that is the way most problems occur.

The text material includes many solved examples. The best way to master the concepts involved in flowcharting is to study each description and the flowchart carefully as it is encountered in the text. Although this will slow the reading somewhat, your comprehension of the concepts will be greatly enhanced.

Acknowledgements go to my wife for having the audacity to argue with me, Weldon Rackley for the push, Robert Cloud and Roland Smith for moral support, and Ellie Perry and Christie Ferrino for typing.

Ronald E. Elliott

Contents

Problem Solving and Flowcharting

INTRODUCTION

The Nature of Problem Solving

A great mystique exists concerning computers and programming. One hears many tales of computer "magic" and of the wild and wonderful feats a computer can perform. These ideas are sustained only because many people are uninformed.

A computer is a tool which merely serves to *assist* in problem solving. As a shovel helps to dig a hole, a computer helps to solve a problem. Computers share a common drawback with many other tools – alone they are useless. The human brain is the only device capable of creative thinking.

Since this is the case, we should make every effort to improve our problem solving ability. The purpose of this chapter is to provide a systematic approach to problem solving which will apply in particular to computer programming problems.

The first step in problem solving is to define the problem. The problem statement should be examined carefully to determine what question must be answered. For example, consider this problem. A farmer had 17 sheep and all but 9 of them died. How many sheep did the farmer have left?

The first impulse is to subtract 9 from 17 to obtain an answer, but careful examination of the problem shows that this method will determine how many sheep died, not how many sheep are left.

Again, study each problem carefully and be sure to determine what question must be answered before attempting to solve the problem.

After the problem is defined, the next step is to assemble all given data and to assign variable names to those quantities which are unknown. Statements such as

Let x = the distance from point A to point B

are the basis of this step.

Next, all unnecessary information should be discarded. Many problems occur containing information which has no meaning to the solution.

Consider this problem.

A man received a check for $88.60 of which he deposited $80 into his checking account. If his account balance was $180 after this deposit, how much was his balance before the deposit?

In this problem, the amount of the check does not matter; only the amount deposited to the checking account affects the problem solution. In this case, the given figure of $88.60 should be discarded.

The next step is to discover relationships among the data and to express these relationships as equations. For example, after reaching the green, a golfer used his putter to measure the distance from his golf ball to the hole. He found the distance to be 5 putter lengths.

Let p = the length of the putter

d = the distance from the ball to the hole

Thus, $d = 5p$ represents the relationship expressed in the problem. The final step of problem solving is to arrange the equations into an algorithm. An *algorithm* is a sequential set of procedures which will produce a desired result. After the algorithm is established, the actual procedures may be executed by either a computer or manually.

The following example will illustrate these steps. A company has 1000 employees of whom 40 percent are male. The company gave a party which 65 percent of the employees attended. If 45 percent of those attending were men, how many women employees attended the party?

Solution:

1. The question is: "How many *women* attended?"
2. Let x = the total number of employees attending

   y = the number of men attending

   z = the number of women attending
3. The given fact that 40 percent of the employees are male has no bearing on the problem and may be discarded.
4. The relationships stated are:

   (a) 65 percent of the employees attended (x)

   (b) 45 percent of those attending were male (y)

   (c) This implies that the difference between x and y represents the number of women attending (z)

Express these relationships as equations to obtain:

(a) $x = (1000)\ (0.65)$

(b) $y = (x)\ (0.45)$

(c) $z = x - y$

5. Since the variable x must be known to find the variable y and both x and y must be known to find z, the equations must be solved in the above order.

Now, the procedures may be executed to obtain the solution.

In summary, the following steps should be applied to each problem:

1. Define the problem.
2. Assemble known quantities and assign variable names.
3. Discard unimportant data.
4. Establish relationships and express them as equations.
5. Determine the proper algorithm by arranging the equations in sequence.

1

What is a Flowchart?

A *flowchart* is simply a graphic method of indicating a proposed or actual solution of a problem. For example, the flowchart shown in Figure 1–1 illustrates a plan to drive to town to do some shopping.

In this example, the rectangular boxes contain procedures. The boxes are connected by lines with arrowheads. The arrowheads indicate the order of execution of the procedures.

The term *flowchart* is generally used in the technical sense to apply to the procedures to be executed by a computer to solve a problem. However, the principle remains the same – the flowchart must show what is to be done and must indicate the sequence.

A correctly drawn flowchart allows the actual computer programming to be accomplished with much greater ease. Therefore, a cardinal rule of good programming technique is "flowchart now, code later." The flowchart helps with the organization of ideas and allows quick recording of the ideas on paper in an easy-to-follow arrangement. Most programmers agree that the most difficult phase of problem solution is determining the desired procedures and arranging them in the

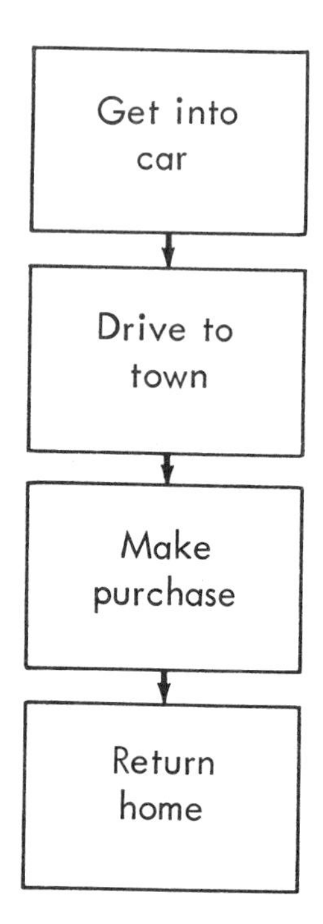

Figure 1-1

proper sequence. Since the flowchart helps with this, once the flowchart is completed the battle is more than half won.

Defining the steps to be used in the solution of a problem is usually an involved process requiring concentrated effort. After this is completed, one may find it difficult to retrace his original chain of thoughts. A good flowchart, drawn when the ideas were fresh, will serve to remind you at a later time what you had in mind at the time the work was begun.

In addition to being useful during coding, the flowchart is an excellent medium to convey ideas to other programmers. To facilitate idea transfers and to avoid confusion, some flowcharting standards should be adopted within each organization. Computer manufacturers supply templates which provide standards. The standard for this text is the IBM template Form X20-8020-1 illustrated in Figure 1–2.

In addition to the use of standard symbols, there must also be a definition of the amount of detail to be given in a particular flowchart. The inclusion of details can easily get out of hand. To illustrate this point, the problem charted in Figure 1–1 could either be expanded as in Figure 1–3 or could be

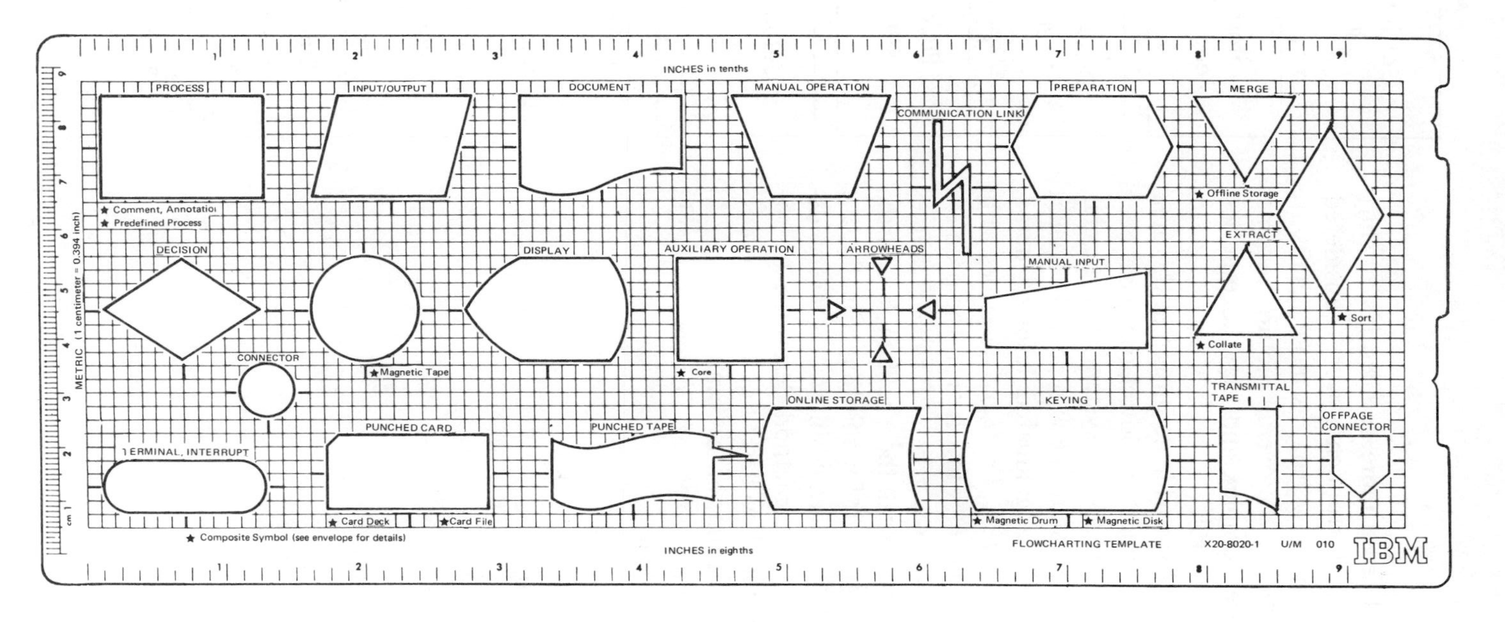

INCHES in tenths
PROCESS
INPUT/OUTPUT
DOCUMENT
MANUAL OPERATION
COMMUNICATION LINK
PREPARATION
MERGE
★ Offline Storage
★ Comment, Annotatio
★ Predefined Process
DECISION
DISPLAY
AUXILIARY OPERATION
ARROWHEADS
MANUAL INPUT
EXTRACT
★ Sort
★ Collate
CONNECTOR
★ Magnetic Tape
★ Core
ONLINE STORAGE
KEYING
TRANSMITTAL TAPE
OFFPAGE CONNECTOR
TERMINAL, INTERRUPT
PUNCHED CARD
PUNCHED TAPE
★ Card Deck
★ Card File
★ Magnetic Drum
★ Magnetic Disk
★ Composite Symbol (see envelope for details)
INCHES in eighths
FLOWCHARTING TEMPLATE
X20-8020-1 U/M 010
IBM
METRIC (1 centimeter = 0.394 inch)

Figure 1-2

made even more detailed if gear shifting and stoplights were considered. In any flowchart, as more details are included, the overall picture becomes more complex and may become less meaningful.

To provide definition in this area, three types of flowcharts have been established: (1) system, (2) general, and (3) detail. Each type of flowchart involves a graduated level of detail. The *system flowchart* describes only the machines involved in the problem solution; the *general flowchart* defines exactly what the problem is and the general plan for solution. Finally, the *detail flowchart* illustrates every detail of the method described by the general flowchart. For complete documentation of a computer program all three levels of flowcharts should be drawn.

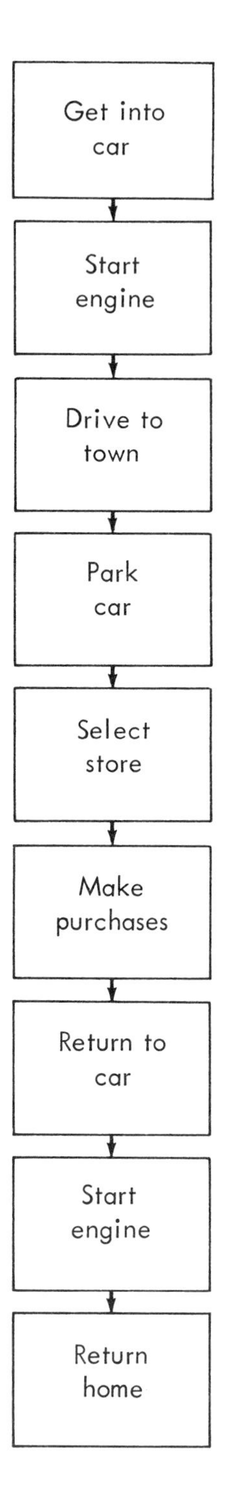

Figure 1-3

EXERCISES

1. Define *flowchart*.

2. Name the three levels of flowcharts.

3. Give a brief description of each level of flowchart.

4. List two good reasons for using flowcharts.

5. Use a block flowchart (as in Fig. 1–3) to illustrate the procedure involved in changing an automobile tire.

2

System Flowcharts

As stated previously, a system flowchart serves to identify the devices to be used in the solution of a particular problem. The most commonly used devices are: tape drives, printers. paper tape readers, console keyboards, and the computer itself. Since the system flowchart is not concerned with the procedures involved, only these input and output devices need to be shown. The flowcharting symbols associated with these devices are illustrated in Figure 2–1. These symbols are connected by lines with arrowheads showing the flow from one symbol to another to create a flowchart. A brief comment may be printed inside each symbol to indicate what the symbol represents. For example, Figure 2–2 shows the process of reading some punched cards, processing the data from the cards, and producing a printed report.

Actually, the symbols are so nearly self-explanatory that the comments are not required. Some programmers prefer to show only the identification of the devices involved. If an IBM 1130 system with an 1132 printer and a 1442 card reader were being used, the flowchart of Figure 2–2 could be drawn as shown in Figure 2–3. In either case, the symbols are constant, and the same general idea is conveyed.

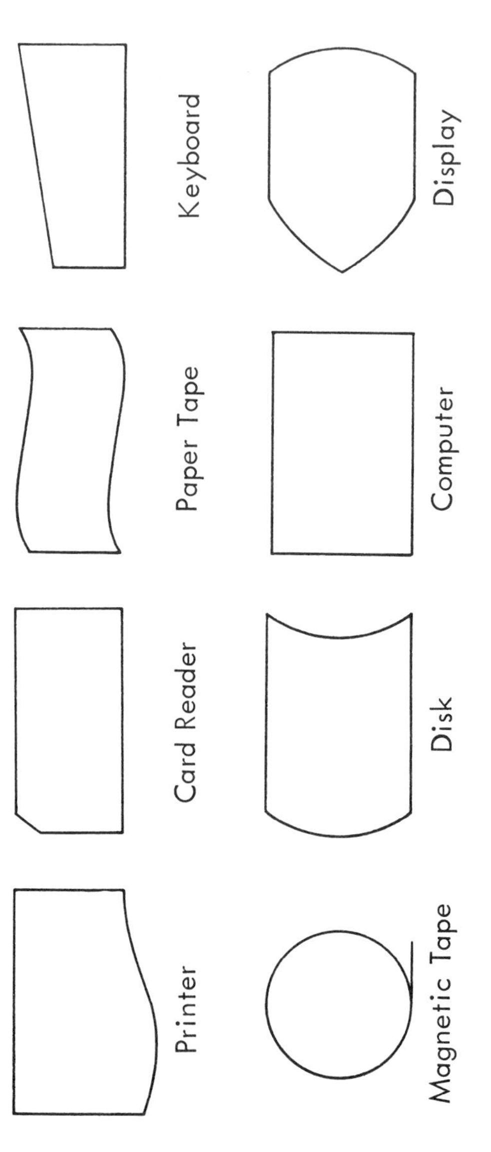
Printer
Card Reader
Paper Tape
Keyboard
Magnetic Tape
Disk
Computer
Display

Figure 2-1

In many cases, the input and/or output of the program will come from several sources rather than from only one source. For example, consider a payroll application where the employee's name, number, and rate of pay are contained on a magnetic tape and his name, number, and hours worked are punched into a time card. A programmer must read the data from both media, compute the amount of the paycheck, and print the paycheck. The system flowchart for the problem appears in Figure 2–4.

For a review of the system flowchart, remember that the programmer makes no attempt to show what the problem is or the logic involved in the solution, only the machines to be used. The proper symbol must be used for each device, and they must appear in the proper sequence. As in all flowcharts, neatness counts, and arrowheads are important because they indicate the direction of flow.

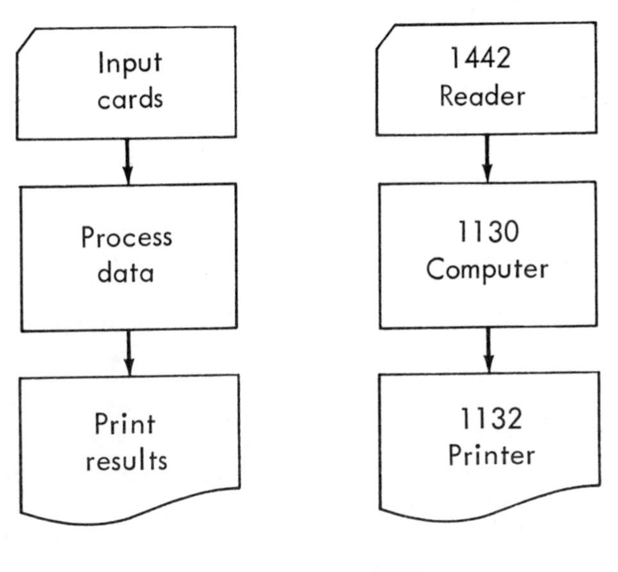

Figure 2-2 **Figure 2-3**

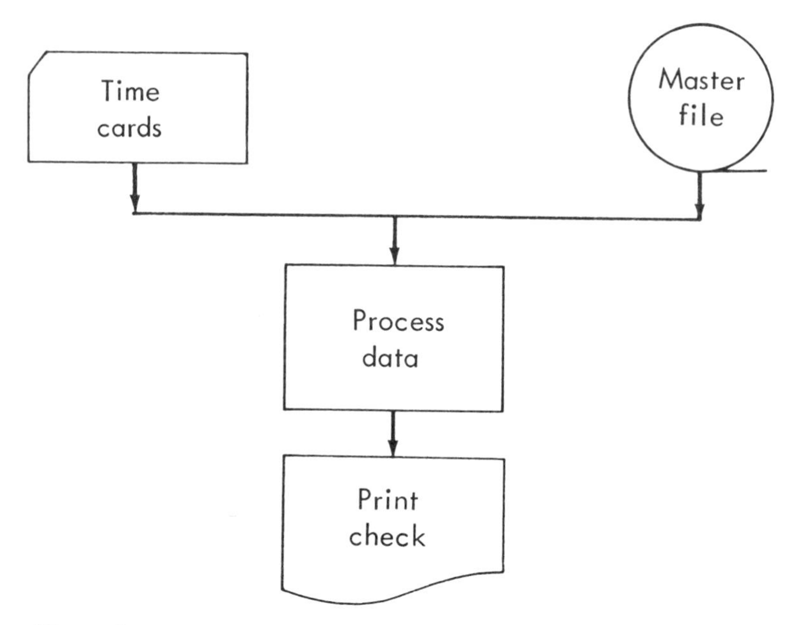

Figure 2-4

EXERCISES

1. Discuss the purpose of system flowcharts.

2. Why are arrowheads used in a flowchart?

3. Draw system flowcharts for the following problems:

   a. A program must read a paper tape, process the data, and produce a printed report.

   b. Data are to be input through a console keyboard, combined with data from a disk, and processed. As output, the disk is to be updated with the results, and the results will also be written onto a magnetic tape.

3

General Flowcharts

The next level of complexity in flowcharting is the general flowchart. A general flowchart is used to show, *in general terms,* what is to be accomplished by the program. When a programmer draws the general flowchart, he begins to organize his thoughts into a logical step-by-step procedure. A programmer must remember that he is dealing with a computer and not with another person. A machine cannot understand any procedure or make any assumptions, the misnomer "electronic brain" notwithstanding. The programmer must show every step of the sequence to produce the desired solution.

In some instances, someone who is not familiar with computer work may be required to understand a program. Therefore, the general flowchart should show, in non-technical language, what the program is designed to do.

Normally, only four symbols will be employed in general flowcharts. These four symbols are shown in Figure 3–1. As in all flowcharts, these symbols are connected by lines with arrowheads to define the direction of flow from step to step.

To draw general flowcharts, one should first become familiar with the usage of each of these symbols. Every flowchart

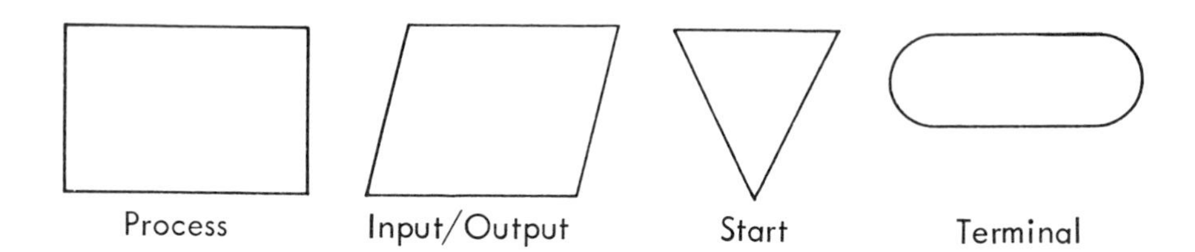

Figure 3-1

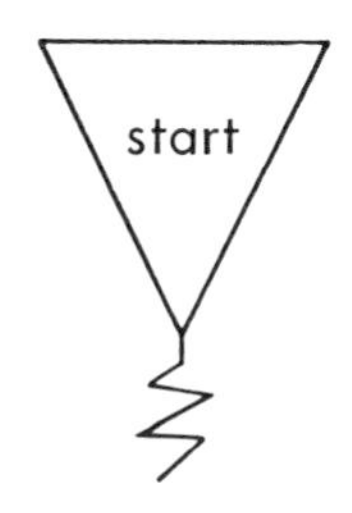

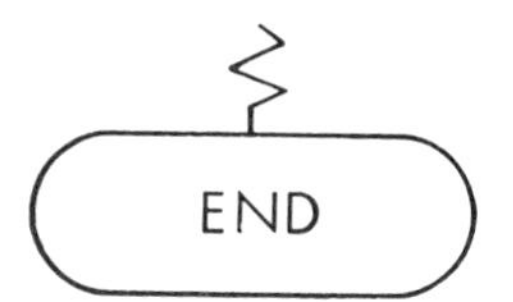

Figure 3-2

should begin with a "start" symbol and end with a "terminal" symbol. The start and terminal symbols usually will appear only once each in a flowchart. The usage of these two symbols is illustrated in Figure 3–2. Notice that the word *start* is written inside the start symbol, and that the word *end* is written inside the terminal symbol. In this usage, the words are not commands to the computer, but merely indicate the physical starting and ending points of the flowchart.

The remainder of the flowchart will be composed of "process" and "input/output" symbols. The next step in completing the flowchart is to mentally analyze the problem in order to obtain a logical sequence of processes. A flowchart usually will begin with an input step, followed by some type of processing, followed by an output step. This sequence occurs so often that it has been named the *data processing cycle*. It appears in Figure 3–3. Most problems are, of course, more complex than this basic form. A problem may have input and output interspersed throughout the processing steps; however, this basic format appears often.

The input/output symbol will appear at any point at which data are to be entered into the computer or any point at which data are to be obtained from the computer. The process symbol is used to indicate the steps involved in manipulating the data into the desired result. Comments to accompany these symbols are essential to explain the procedures. Many programmers write inside the symbols, but this practice may lead to cramped, illegible writing. Therefore, some programmers prefer to write comments outside the symbols. Either practice is acceptable, provided that the neatness of the flowchart does not suffer and the procedure is consistent within each flowchart.

Consider the problem of Figure 2–4. The general flowchart associated with this problem is shown in Figure 3–4. This flowchart illustrates two important points. First, in the input/output symbols, it is not necessary to indicate the devices involved. Secondly, the process block gives no clue as to *how* the pay is to be computed. It only shows the *general* fact that the computer must calculate the amount of pay. Also, the data processing cycle may be detected in Figure 3–4.

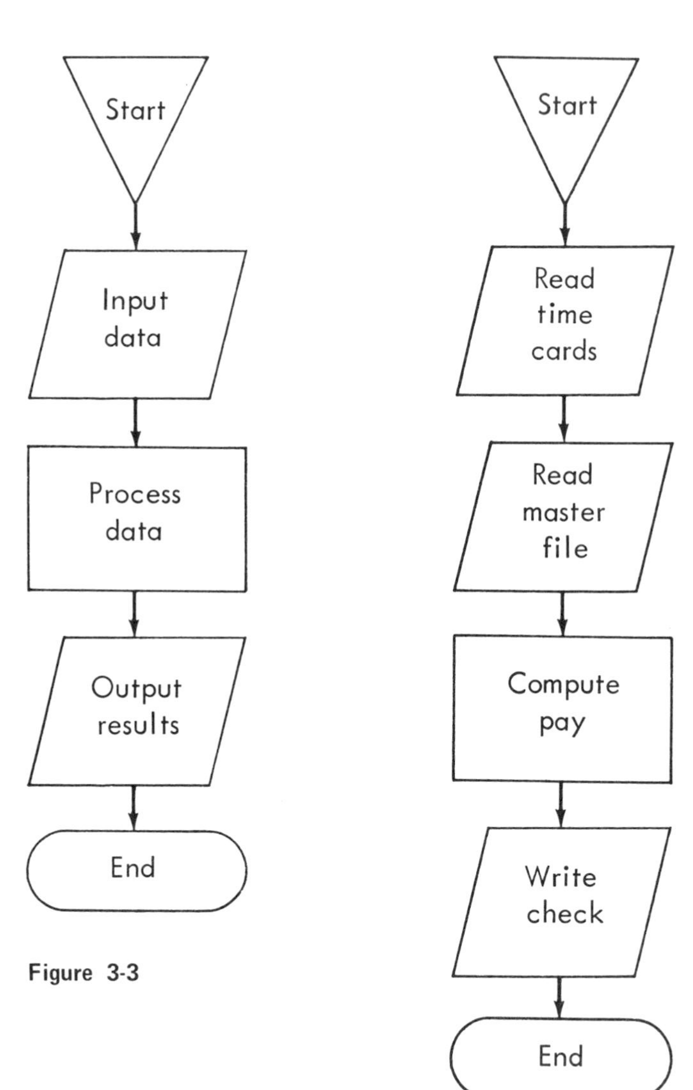

Figure 3-3

Figure 3-4

A company maintains two magnetic tapes that describe their stock of merchandise. One tape contains the item stock number and the amount of stock on hand for that item. The other tape contains the same item stock number and the item description. In addition, the company provides cards that are punched with the same item stock number and the number of items sold. A program is required to read all of this data, com-

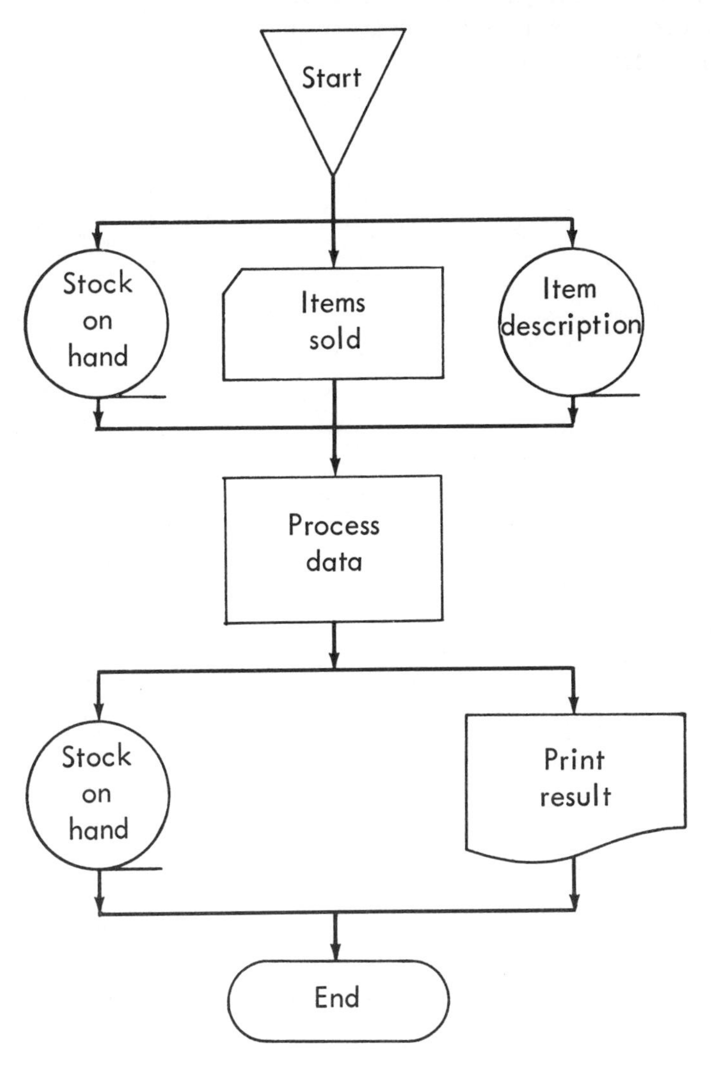

Figure 3-5

pute the current stock on hand, print the result, and update the stock-on-hand tape. The system flowchart for this problem appears in Figure 3–5. The associated general flowchart is shown in Figure 3–6. The student should compare these two flowcharts to cement the understanding of the two different types.

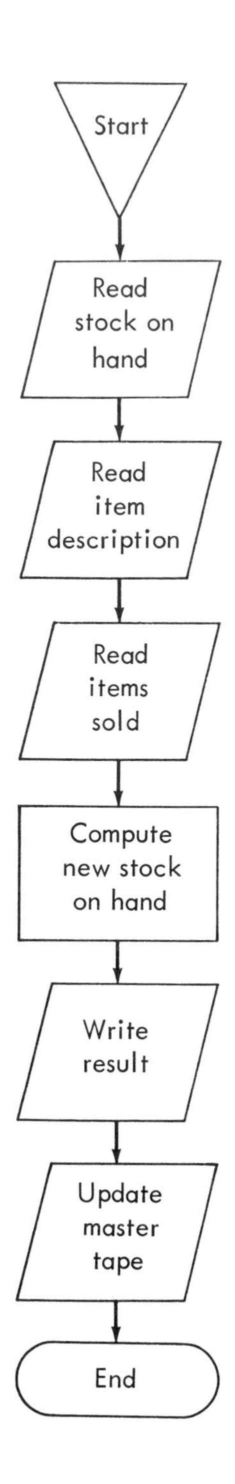

Figure 3-6

EXERCISES

1. Why are comments necessary to accompany the symbols in general flowcharts?

2. Why is it not necessary to specify a device for the input/output symbol?

3. In a problem such as that shown in Figure 3–4, where would you expect to find the steps that show how the pay is computed?

4. Draw the system and general flowcharts for the following problem:

   A bank maintains a disk file that consists of an account number and an account balance. A card file is provided that indicates the withdrawal and deposit totals. The program is to read this data, compute the current balance, print a report, and update the disk file.

5. A company uses a computer to print form letters from the appropriate department. The name and address are to be read from a tape file, the letter text printed, and the appropriate department manager's name selected from a disk file and printed at the bottom of the letter. Draw the system and general flowcharts to illustrate this procedure.

4

Detail Flowcharts

A mathematician has been employed by a company to compute the total amount of pay for each of the employees. He is given the hourly rate of pay of each employee, the total number of hours worked by each, and a helper who can perform simple arithmetical computations but has no other mathematical knowledge. The mathematician's problem is to develop a set of procedures which will direct his helper to the solution. The method he must follow is to analyze the solution and to break it down into simple steps which the helper can perform.

This is essentially the situation a programmer faces when he attempts to program a computer. A computer cannot carry on a conversation and is not capable of any mental process. Basically, a computer can perform only simple arithmetical functions and can store the result. Therefore, the programmer must supply a series of logical steps in order to produce the desired solution. The steps must consist of nothing more than addition, subtraction, multiplication, and division.

Placing operations on such a simple basis may be difficult for many people. To illustrate, the solution to the above problem is obviously to multiply the hours worked by the

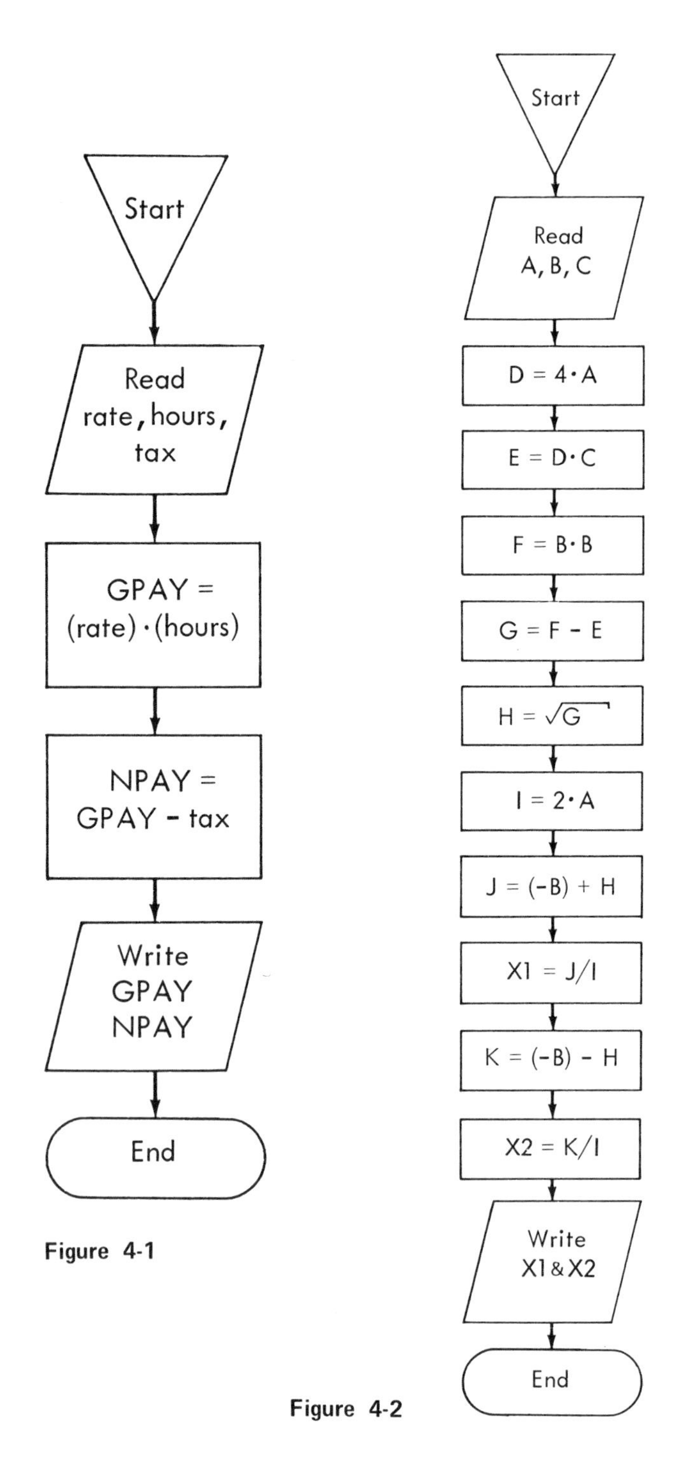

Figure 4-1

Figure 4-2

rate per hour to obtain the gross pay. However, this solution is not obvious to a computer. If the problem is expanded to include the deduction of taxes, a typical programming situation is created. The programmer's analysis might go like this:

1. Input the hourly rate, the hours worked, and the tax deduction.
2. Multiply the hourly rate by the hours worked, giving the gross pay.
3. Subtract the tax deductions from the gross pay, giving the net pay.
4. Output the gross pay and the net pay.

This list of steps could be followed easily by the helper. Programmers could show their analyses of all problems in this manner, but it is much easier and more concise to draw a detail flowchart. By using the same symbols and logic as used in general flowcharts, the solution of this problem can be shown as in Figure 4–1. This method of illustrating the steps involved in the problem solution is easier to write and allows the sequence of operations to be observed at a glance.

The method of solution is not always so obvious as it is in the payroll problem. Consider having a computer solve a quadratic equation of the form $\mathbf{A}x^2 = \mathbf{B}x + \mathbf{C} = 0$.

The programmer's task is to investigate methods of solution and determine the most efficient one. In some cases, we can find the solution to a quadratic equation by inspection. For example, it is apparent that $x = \pm 1$ is the solution to the equation $x^2 - 1 = 0$. We must remember, however, that a computer cannot think and, therefore, cannot possibly find a solution by inspection.

Another method which might be used is the process of factoring the equation into two simpler equations. When the equation $x^2 + 6x + 8 = 0$ is written as $(x + 4)(x + 2) = 0$ the solutions $x = -4$ and $x = -2$ readily follow. The drawback to this method is that it would be very difficult to instruct the computer to transform an equation into its factors. So this method must be rejected.

A third possibility is to complete the square, which involves adding a term to each side of the equation to cause it to factor more easily. However, when this is done, the problem of determining the factors still exists, so this method must also be rejected.

The quadratic formula method, which defines the solution to be

$$x = \frac{-B \pm \sqrt{B^2 - 4AC}}{2A}$$

is the only method which will suffice because it is the method involving only arithmetical computations.

After the method of solution is determined, the programmer must list the steps of implementing that method.

The steps to solve the quadratic formula are as follows:

1. Input **A**, **B**, and **C**.
2. Multiply 4 by **A** and save the result as **D**.
3. Multiply **D** by **C** and save the result as **E**.
4. Multiply **B** by **B** and save the result as **F**.
5. Subtract **E** from **F** and save the result as **G**.
6. Compute the square root of **G** and save the result as **H**.
7. Multiply **A** by 2 and save the result as **I**.
8. Add **H** to the negative of **B** and save the result as **J**.
9. Divide **J** by **I**, producing one root (**X1**).

10. Subtract **H** from the negative of **B** and save the result as **K**.

11. Divide **K** by **I**, producing the second root (**X2**).

12. Output **X1** and **X2**.

Compare this list of steps with Figure 4–2. The flowchart in Figure 4–2 is easier to write and much easier to understand.

If the programmer feels that any possibility exists of a reader misunderstanding any part of the flowchart, an *annotation* box may be included at any point. An annotation box is a rectangular symbol attached to the flowchart by a dashed line. The annotation symbol may contain any additional descriptions or comments the programmer feels are necessary to explain the process being flowcharted.

The same process could be shown in many different ways. The adage, "Give 50 programmers the same problem to flowchart, and the result will be 50 different flowcharts," implies that there is no one "right" flowchart for a problem. However, some elements of all flowcharts should remain constant to facilitate communication between programmers. Therefore, a need exists to establish a few formal rules for drawing flowcharts. Adhering to the following rules, one is considered to be using good flowcharting technique.

1. The direction of flow should usually be from the top to the bottom of the page and from the left to the right.

2. Lines should never cross. Crossing lines can always be avoided by the use of a pair of connector symbols.

3. Arrowheads should be used to show the direction of the flow, especially if it is other than top to bottom or left to right.

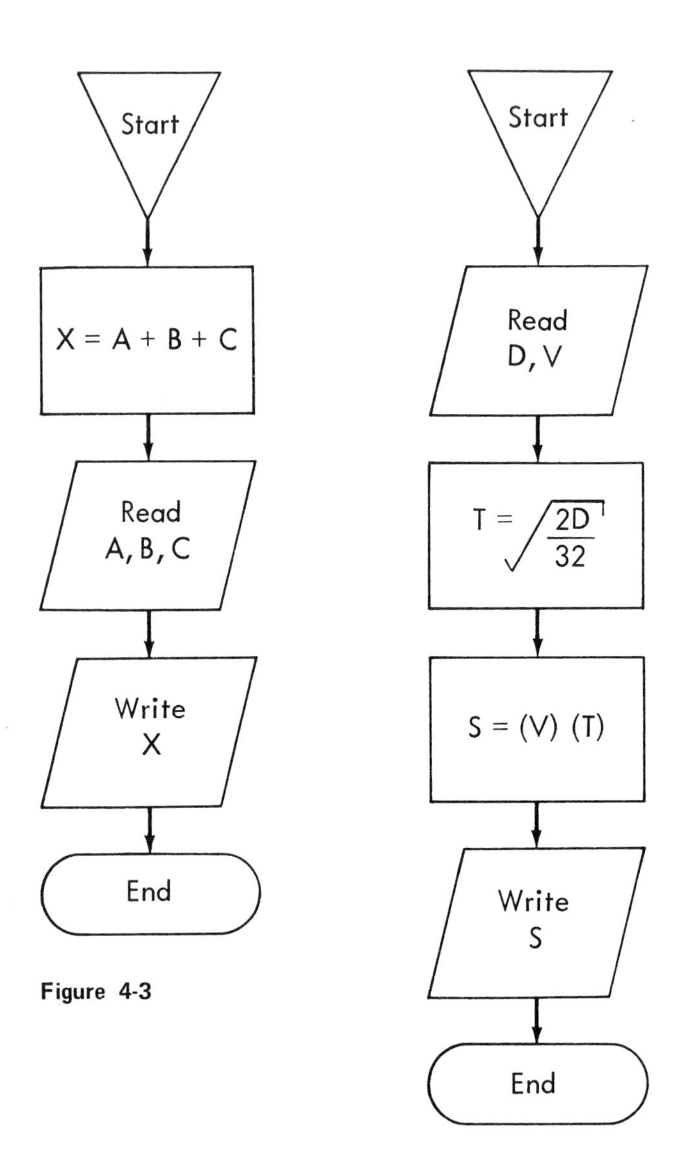

Figure 4-3

Figure 4-4

4. The symbols and lines should be drawn neatly.

5. The symbols may be drawn any size – only the shape is standard.

The student should not be mislead by the simplicity of these rules. The drawing of detail flowcharts is not mastered easily by a beginner. Drawing good flowcharts requires much practice, plenty of paper, and a large eraser.

Beginners often have difficulty with the proper sequence of the steps. Figure 4–3 is an illustration of a common error. The programmer must remember that data cannot be used for computation until after it has been read by some device.

Some programmers like to fill the symbols with comments; other programmers prefer merely to write the equations. Either procedure is acceptable, provided the meaning is clear. If the shorthand method of writing equations is used, a table explaining the variable names should be included.

Consider a final example of problem analysis and of drawing the flowchart.

A bomber with an onboard computer has a faulty bombsight. A program is needed which will give the proper distance from the target at which release of the bomb will produce a hit.

Since the bomb will be affected only by gravity and by the speed of the plane, this program can be written if the altitude and speed are known.

The first step is to determine the time it will take the bomb to fall to the earth by using the gravity formula

$$d = \frac{1}{2} g t^2$$

where d is the altitude, g is the acceleration of gravity (32 ft/sec/sec), and is the time involved. Solve the equation for t.

$$t = \sqrt{\frac{2d}{g}}$$

The second step is to determine how far the bomb will travel forward during this time by using the formula

$$s = vt$$

where s is the distance, v is the velocity of the plane (and bomb), and t is the time found above.
Thus

$$s = v\sqrt{\frac{2d}{g}}$$

The flowchart for this problem is shown in Figure 4–4. In this flowchart, the variable **S** will be the distance the plane should be from the target when the bomb is released. As in all programs, the flowchart is designed to produce the solution for any input values of **D** and **V**.

EXERCISES

1. Why is there no one "right" way to draw a flowchart?

2. How can a programmer determine if his flowchart is correct?

3. Draw detail flowcharts for the following problems:

   a. Use the formula $I = prt$ to compute the interest on a loan of p dollars at r percent for t years.

   b. Read a punched card containing an employee's regular rate of pay, regular hours worked, overtime hours worked, and tax deductions. Given that overtime pay is one and one-half times the regular rate, compute and print the employee's net pay.

   c. Given that a projectile follows the path described by the function

   $$f(t) = 8t - t^2$$

   compute and output the altitude of the projectile at time t.

   d. A company pays its salesmen an 8-percent commission on all sales. Compute the salesman's commission and the customer's cost on a sale of amount A. The customer's cost will include the salesman's commission.

   e. Write a program using the formula

   $$V = 4/3\,\pi r^3$$

   to compute the volume of a sphere with any given radius r.

f. A laboratory technician has only a Fahrenheit thermometer, but he wishes to have temperatures in Centigrade and Kelvin scales also. By using the formulae

$$C = 5/9\,(F - 32)$$

and

$$K = C - 273$$

compute the Centigrade and Kelvin equivalents of any Fahrenheit temperature.

g. A man knows that the gasoline tank of his car holds 24 gallons and that his car averages 18.3 miles per gallon. He starts with a full tank and drives at 45 miles per hour. Find the volume of gasoline left in the tank at any time

h. A baseball catcher has done some research to cut down on base stealing. He has determined that the average runner's speed is 28 ft/sec, and that he can throw the ball at an average speed of 90 mph. If the bases are 90 feet apart, and the catcher has the ball ready to throw, what is the maximum distance from first base he can allow a runner to start in order to throw him out at second base?

5

Questions Please

A programmer would expect most flowcharts to be more complex than those previously discussed. The basic contributor to flowcharting complexity is the fact that questions must be answered during the processing; that is, decisions must be made. In many instances, the computer must choose between two logic paths. Because a computer cannot think, it cannot make any decision. Therefore, the programmer must reduce the required decision to a comparison of two values. Such a comparison is actually a subtraction which the computer can perform. Because decisions or comparisons occur so frequently, a special method exists to flowchart the process. The symbol shown in Figure 5–1 is a decision box. The appearance of this symbol in a flowchart indicates that a question is being asked. The question is written inside the decision symbol, and the direction of flow from that point will depend on the answer to the question. The decision box must have *at least* two paths leading from it. Each path is labeled to indicate the condition which will cause the flow to follow that path.

The following symbols are used:

Symbol	Definition
$:$	Is compared to
$>$	Is greater than
$<$	Is less than
$=$	Is equal to
$\neq$	Is not equal to
$\geq$	Is greater than or equal to
$\leq$	Is less than or equal to

A decision box in which the value A is compared to the value B is shown in Figure 5–2. The labels indicate that the path to the right is to be followed if A is greater than or equal to B; the path downward is to be followed if A is less than B.

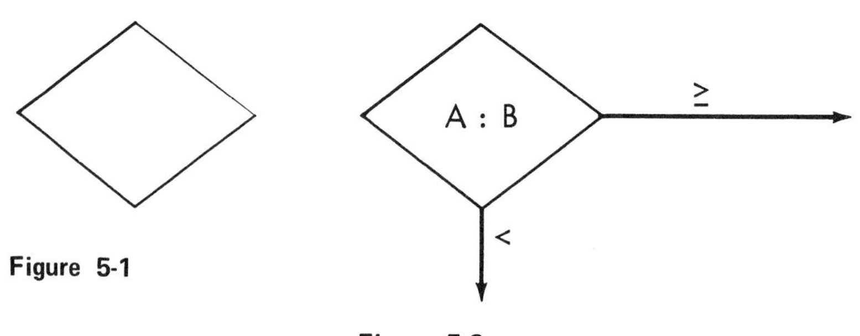

Figure 5-1

Figure 5-2

At first, it may be easier to state each decision such that it will have exactly two answers. This can usually be done by making the decision a question rather than a comparison. For example, the decision in Figure 5–4 could be worded: "Is C *equal* to zero?" Then, the paths leading from the decision box

would be labeled "yes" and "no." The decision shown in Figure 5–6 could be stated as: "Is SALE greater than 100?", and the paths again labeled as "yes" and "no."

A problem requires the computer to read the values A, B, and C and to compute the value of $\frac{A+B}{C}$ as output. The flowchart is shown in Figure 5–3. To be correct, Figure 5–3 should be modified to allow for the possibility of the value C being equal to zero. Figure 5–4 investigates this possibility and omits the division if C is equal to zero.

A common error made by beginners is shown in Figure 5–5. The decision box shown directs the computer to follow the same path regardless of the outcome of the comparison. If this is the desired procedure, then the comparison is unnecessary. Remember that at least two different paths must lead from the decision box, and that they must lead to different points in the flowchart.

A department store pays its salesmen a 5-percent commission on sales of $100 or less and an 8-percent commission on sales of more than $100. A computer must read an amount of a sale and compute the salesman's commission. The flowchart is shown in Figure 5–6.

After the sales amount SALE is read, it is compared with $100. If SALE is less than or equal to $100, the equation C = (SALE) (0.05) is used to compute the salesman's commission (C). If SALE is greater than $100, the other path leading the equation C = (SALE) (0.08) is used to compute the commission. In either case, the value of C is the desired output. When a decision box is included in a flowchart such as Figure 5–6, the necessity of arrowheads is evident.

Some situations require several decisions in series. Consider a problem in which a card contains the four values A, B, C and CODE. The problem states the following conditions:

X = A + B + C if the CODE is 0

X = A + B if the CODE is 1

X = A + C if the CODE is 2

X = A – C if the CODE is 3.

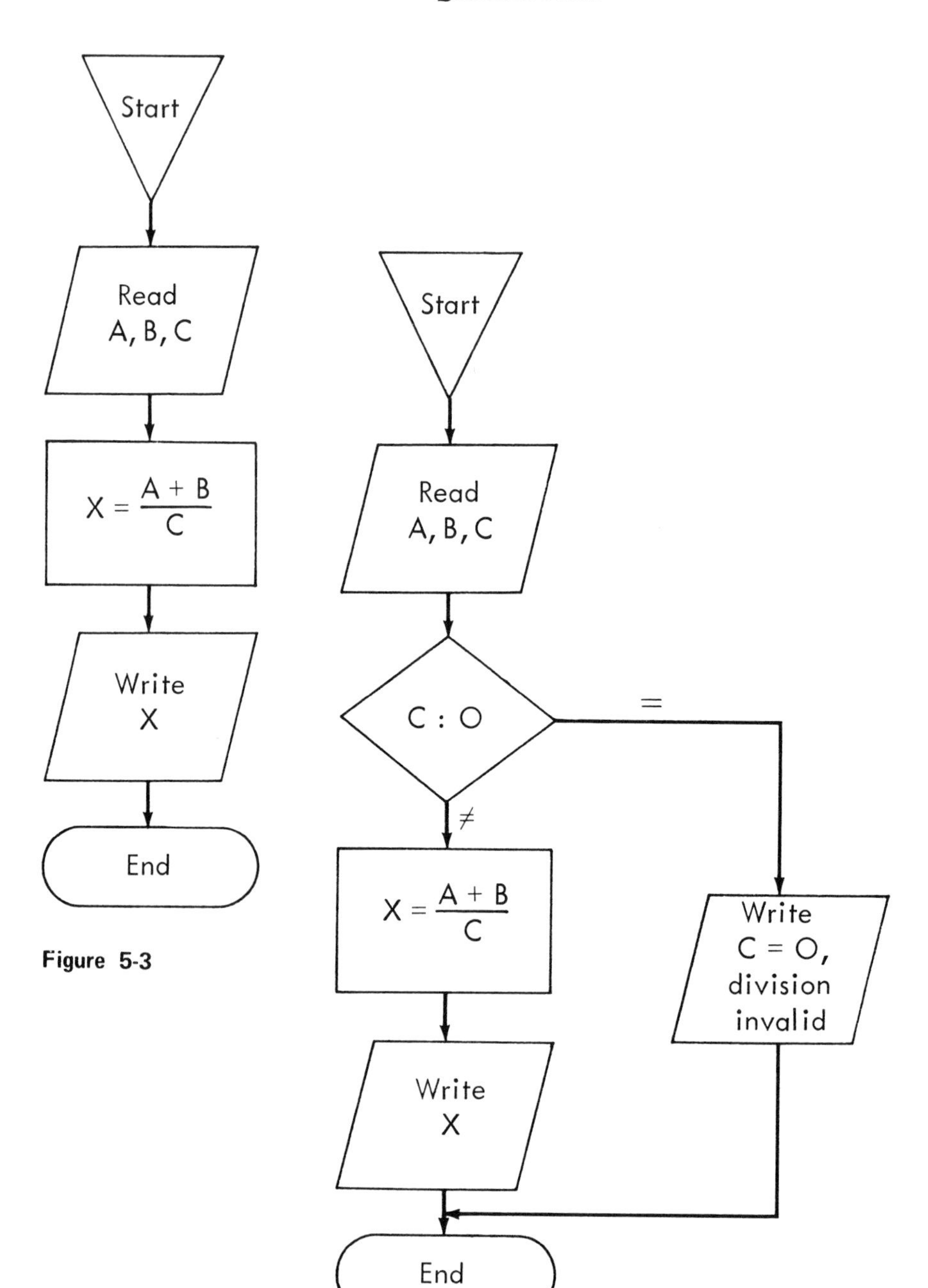

Figure 5-3

Figure 5-4

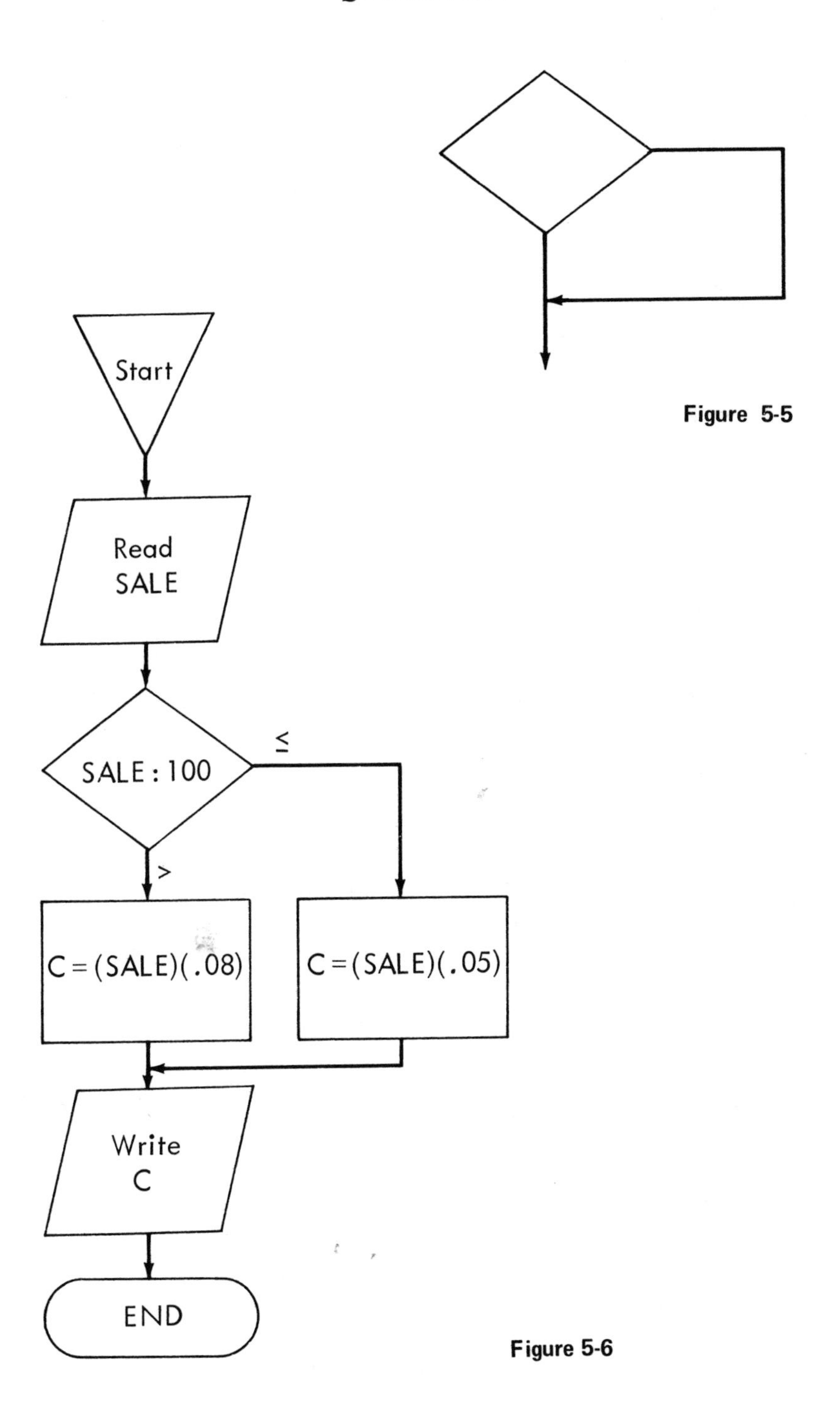

Figure 5-5

Figure 5-6

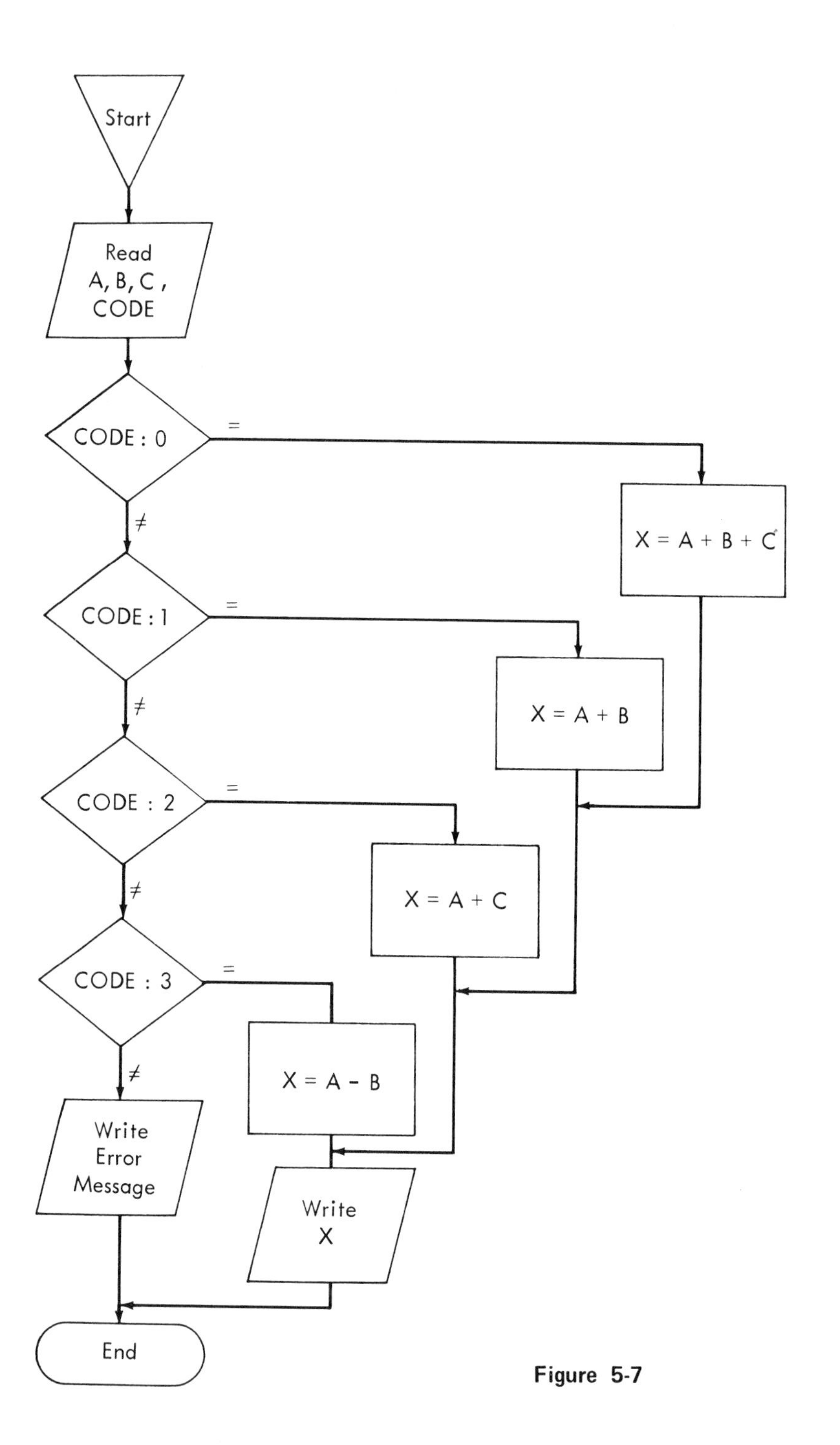
Start
Read
A, B, C ,
CODE
CODE : 0
=
≠
X = A + B + C
CODE : 1
=
≠
X = A + B
CODE : 2
=
≠
X = A + C
CODE : 3
=
≠
X = A - B
Write
Error
Message
Write
X
End

Figure 5-7

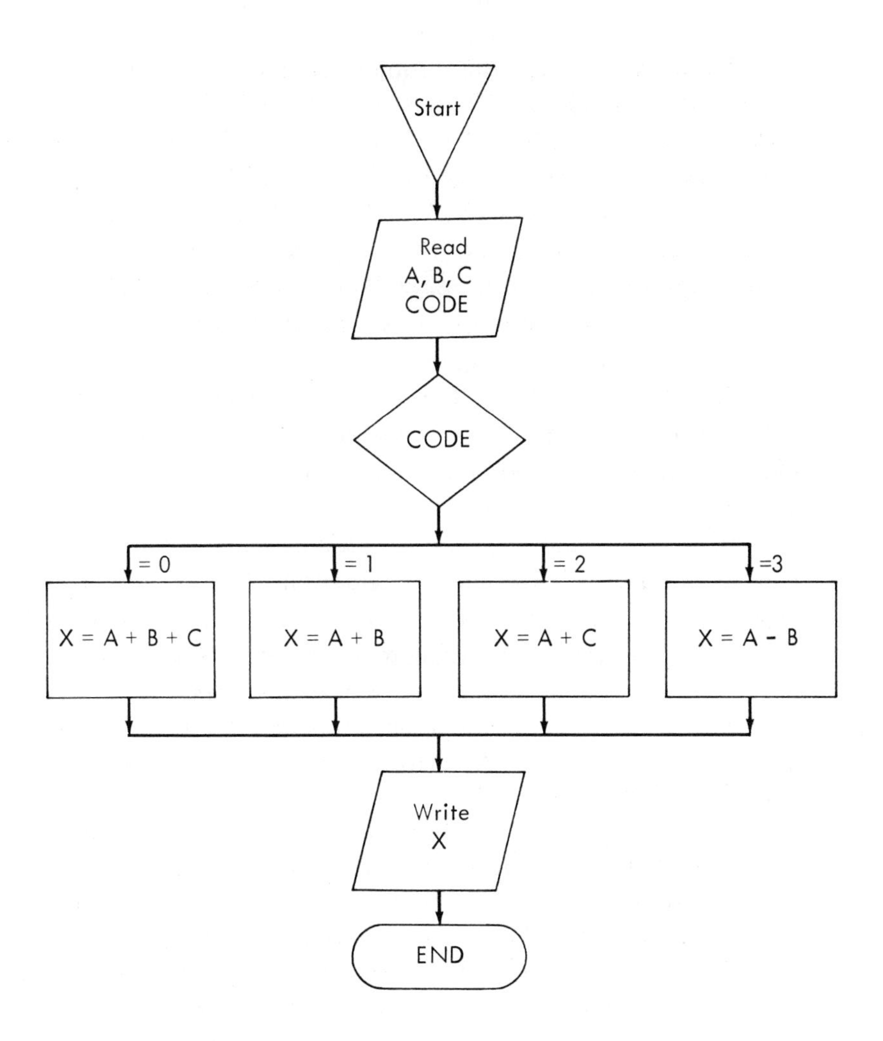

Figure 5-8

The flowchart for this problem (Figure 5–7) requires a series of decisions. First, CODE is compared with zero. If CODE does not equal zero, then it is compared with 1. If CODE does not equal 1, then it is compared with 2. If CODE does not equal 2, then it is compared with 3. If CODE does not equal 3, an error condition exists, and the program will be terminated with an appropriate error message. When the value which CODE equals is found, the appropriate equation is used to compute **X**, and that value is output.

Another method of flowcharting this same problem is shown in Figure 5–8. This method involves several paths leading from the decision box and may be used if the programmer desires.

In some circumstances, a programmer may find it difficult to avoid crossing a line or to prevent drawing a long or jagged line between symbols. The *connector* symbol shown in Figure 5–9 provides a solution to both cases. The connector symbol indicates a transfer of flow and, thus, always must appear in pairs. A character must be written inside the connector so that the counterpart to that connector can be located.

Figure 5-9

An arrowhead leading into a connector indicates that control is to be transferred to the point at which that connector's counterpart appears. An arrowhead leading from a connector indicates the point at which control is to reenter the flowchart. Figure 5–6 could be drawn by using connectors as shown in Figure 5–10. The use of connectors is seldom mandatory but usually will result in a neater flowchart and will allow more flexibility in the flowcharting format.

A note on connector technique – Figure 5–11 illustrates an alternate method of indicating the reentry point. Some programmers prefer this method because it allows the main flow to continue *down* the page. Either method is acceptable.

Many times a transfer will be made from one page of the flowchart to another. When this is the case, the *offpage connector* (Fig. 5–12) is used. Two numbers, separated by a decimal, are written inside the symbol. The first number indicates the page from which the transfer was made; the second

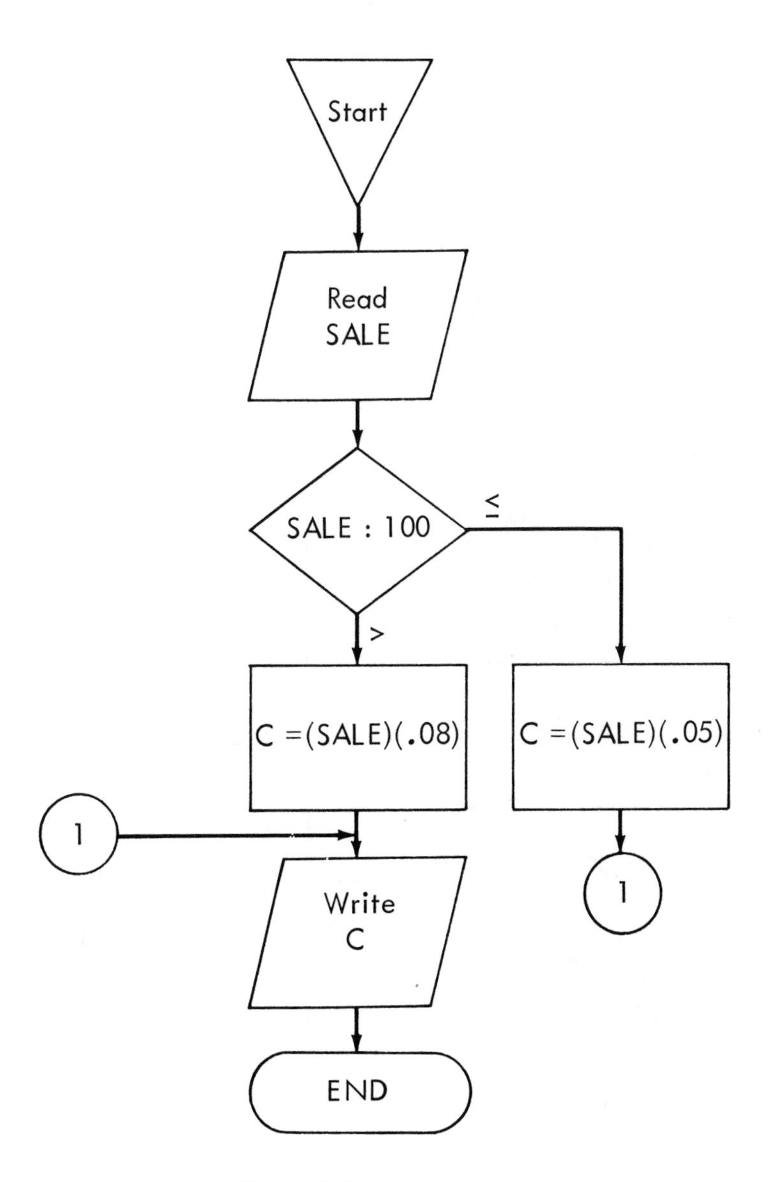

Figure 5-10

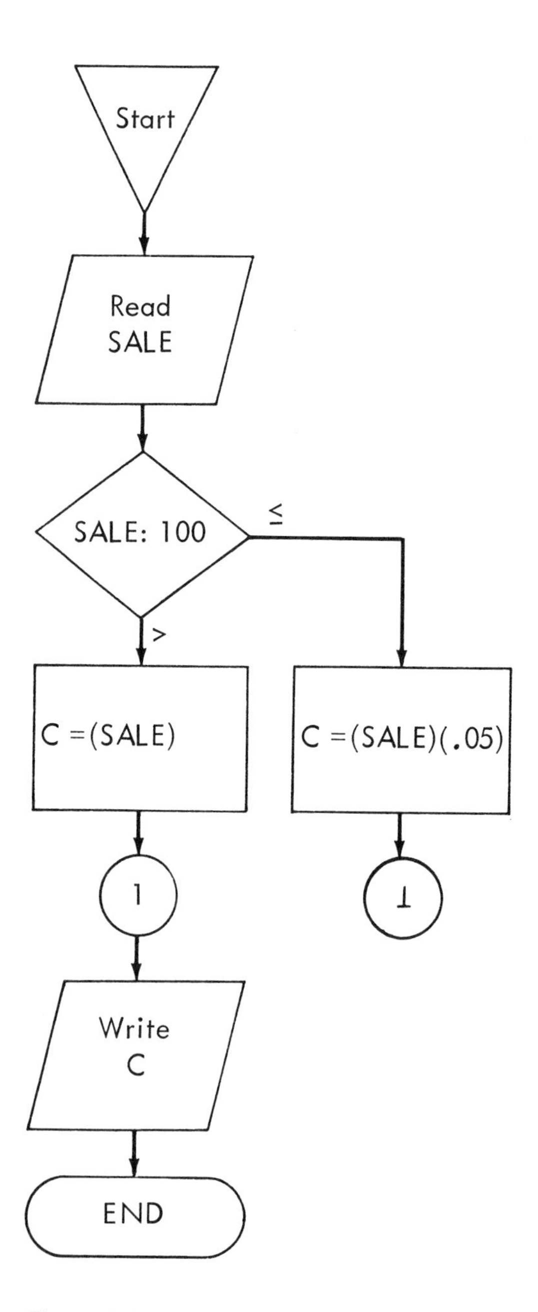

Figure 5-11

number identifies the page on which the reentry point is located. The symbol in Figure 5–12 indicates a transfer from page 1 to page 6. An identical symbol must appear at the reentry point. If more than one offpage connector is required on a page, each offpage connector should include an alphabetical character in addition to the numerical symbols to identify it as shown in Figure 5–13. In a flowchart consisting of several pages, this nomenclature will prevent a reader from having to search for the reentry point.

Figure 5-12 **Figure 5-13**

As a final example, consider the logic involved in finding the largest of three given values. The flowchart is shown in Figure 5–14. The process involves comparing two of the values, discarding the smaller, and then comparing the larger of the first two with the third value. The larger of the last two values compared is the desired answer.

Figure 5–14 illustrates several good flowcharting techniques. Notice that all the paths from the output symbols connect to a single line to enter the terminal. This technique avoids duplicate (or triplicate) terminals.

Also, notice that the connectors used are not really necessary. A line could be drawn which would eliminate the connectors. In this case, the connectors avoid drawing a jagged line.

Two other methods of solving this same problem appear in Figures 5–15 and 5–16. Figure 5–15 shows a roundabout method involving unnecessary computations. This method is poor because the subtracting of **B** from **A** and then comparing the result with 0 produces the same effect as comparing **A** with **B**.

Figure 5–16 shows a more efficient method of solving this problem. This method involves using another variable, called BIG, to hold the largest value. The value of A is initially assigned to BIG, then BIG is compared with B. If B is larger, then the value of B is assigned to BIG. At this point, BIG holds the larger of A and B. The process is then repeated, comparing BIG with C. Regardless of the outcome, BIG contains the desired value for output.

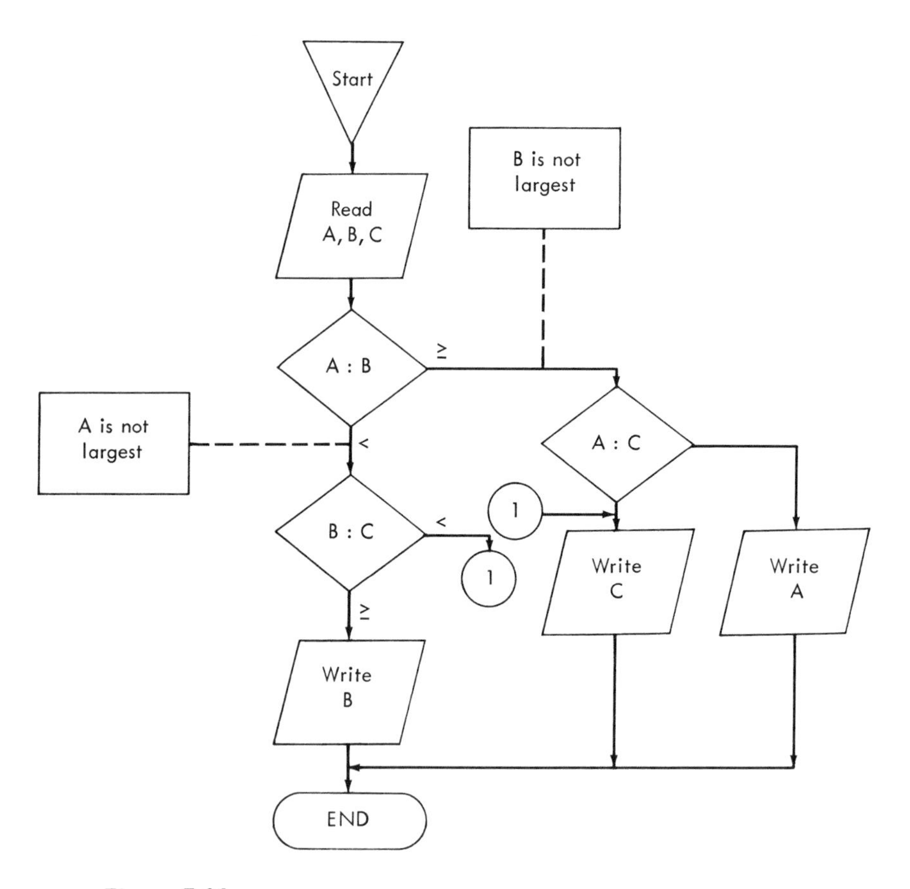

Figure 5-14

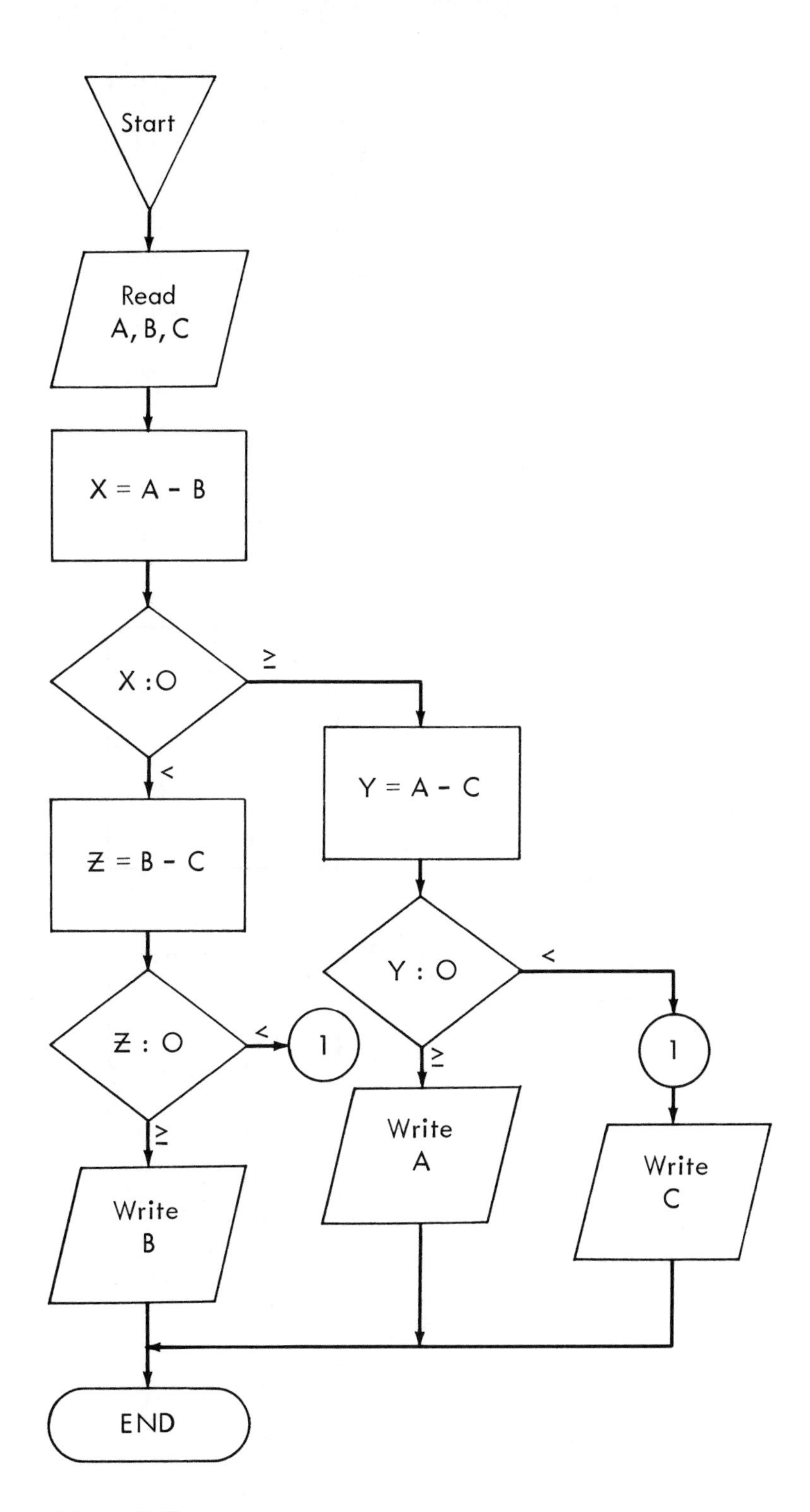
Start
Read
A, B, C
X = A - B
X : O
≥
<
Y = A - C
Z = B - C
Y : O
<
Z : O
<
1
≥
1
≥
Write
A
Write
C
Write
B
END

Figure 5-15

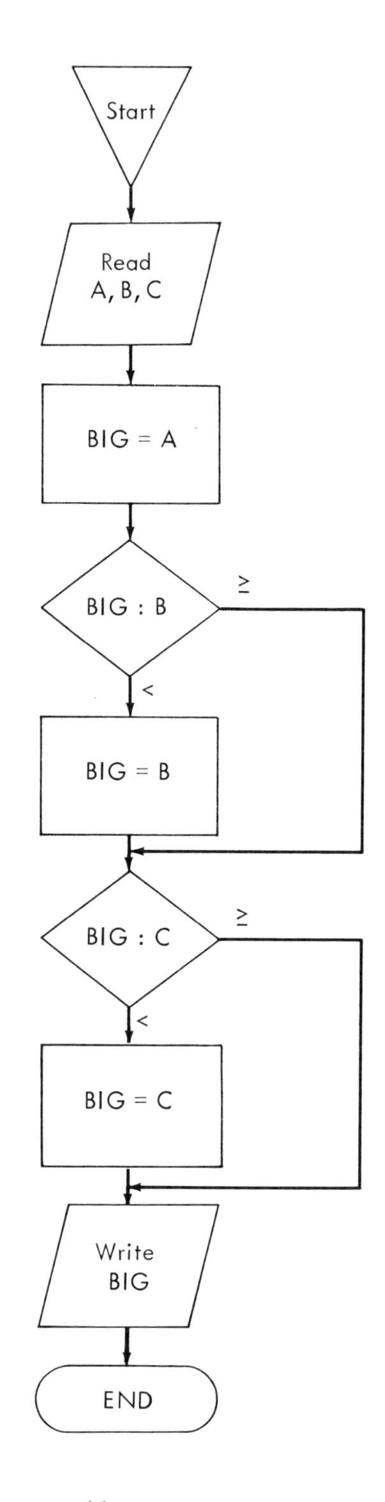

Figure 5-16

EXERCISES

Draw flowcharts for the following problems:

1. Show the logic involved in the decision to reorder an item when (a) the stock on hand, plus that on order, is below the reorder point, and (b) the item is not obsolete.

2. Compute the cash value of a $25 Savings Bond at any time according to the following:

   The purchase price is $18.75. If the bond is less than 6 months old, the interest is zero percent. If the bond is between 6 and 12 months old, the interest rate is 1 percent. If the bond is between 12 and 18 months old, the interest rate is 2 percent. If the bond is between 18 and 24 months old, the interest rate is 3.5 percent.

3. A stock pays a dividend of 8.5 percent to holders of 100 or more shares and 8.0 percent to holders of less than 100 shares. Read a number of shares from a card and compute the dividend, assuming that the selling price is $280 per share.

4. A company classifies its products by weight as follows:

   Class A – 12 pounds or over

   Class B – 5 to 11.9 pounds

   Class C – less than 5 pounds

   Show the logic of determining into which class to place an item.

5. Show the solution of a quadratic equation of the form

   $$AX^2 + BX + C = 0.$$

   Take into account the fact that a computer cannot determine the square root of a negative number.

6. An electric company bases its charges on two rates. Customers are charged $0.0134 per kilowatt-hour for the first 300 kilowatt-hours used in a month and $0.036 each for all kilowatt-hours used thereafter. Compute the amount due from a customer after reading the kilowatt-hours used.

7. A new sales tax went into effect on January 1. Read a card containing the amount and the date of a sale. Compute the amount of the sales tax according to the following schedules.

   Old method (prior to Jan. 1):
   Tax = 3 percent of the sales amount.

   New method (after Jan. 1):

Sales amount	Tax
$ 0.00 to $ 0.99	$ 0.04
$ 1.00 to $ 1.50	$ 0.05
$ 1.51 to $ 5.00	$ 0.10
$ 5.01 to $10.00	$ 0.15
Above $10.00	3.5 percent of sales amount.

8. A company uses timeclocks which print the time in 24-hour clock fashion (i.e., 0100 = 1:00 a.m., 1300 = 1:00 p.m., and so forth). Read a card containing an "in" time and an "out" time and compute in minutes the difference between the two times. The times are less than 24 hours apart, and the first time is earlier in the 24-hour period than the second time. (Caution: the times are not necessarily in the same day.)

9. Read a card containing a date punched as month, day, year and compute the sequential day of the year for that date. For example, 5/14/71 was the 134th day of that year. Take leap year into account.

6

Looping and Counting

Economically, the cost of writing a computer program to perform a process only once cannot be justified. The most efficient use of a computer is performing tasks which require repetition. Consequently, most programs are designed to process multiple sets of input data. This situation requires the programmer to instruct the computer to repeat the processing until all input data have been processed.

The flowchart shown in Figure 6–1 instructs the computer to read three values (A, B and C) and to compute the sum of the first two values divided by the third value. Before the computation is performed, the value C is tested and the computation is omitted if C is equal to 0. If C does not equal 0, the computation is performed and the result is output. In either case, the flowchart tests the input record to determine if it was the last input record. If so, the processing is terminated. If the input record was not the last record, the flowchart "loops" back to the input step, and the entire process is repeated with new values for A, B, and C. Processing continues in this manner until the last input record is processed.

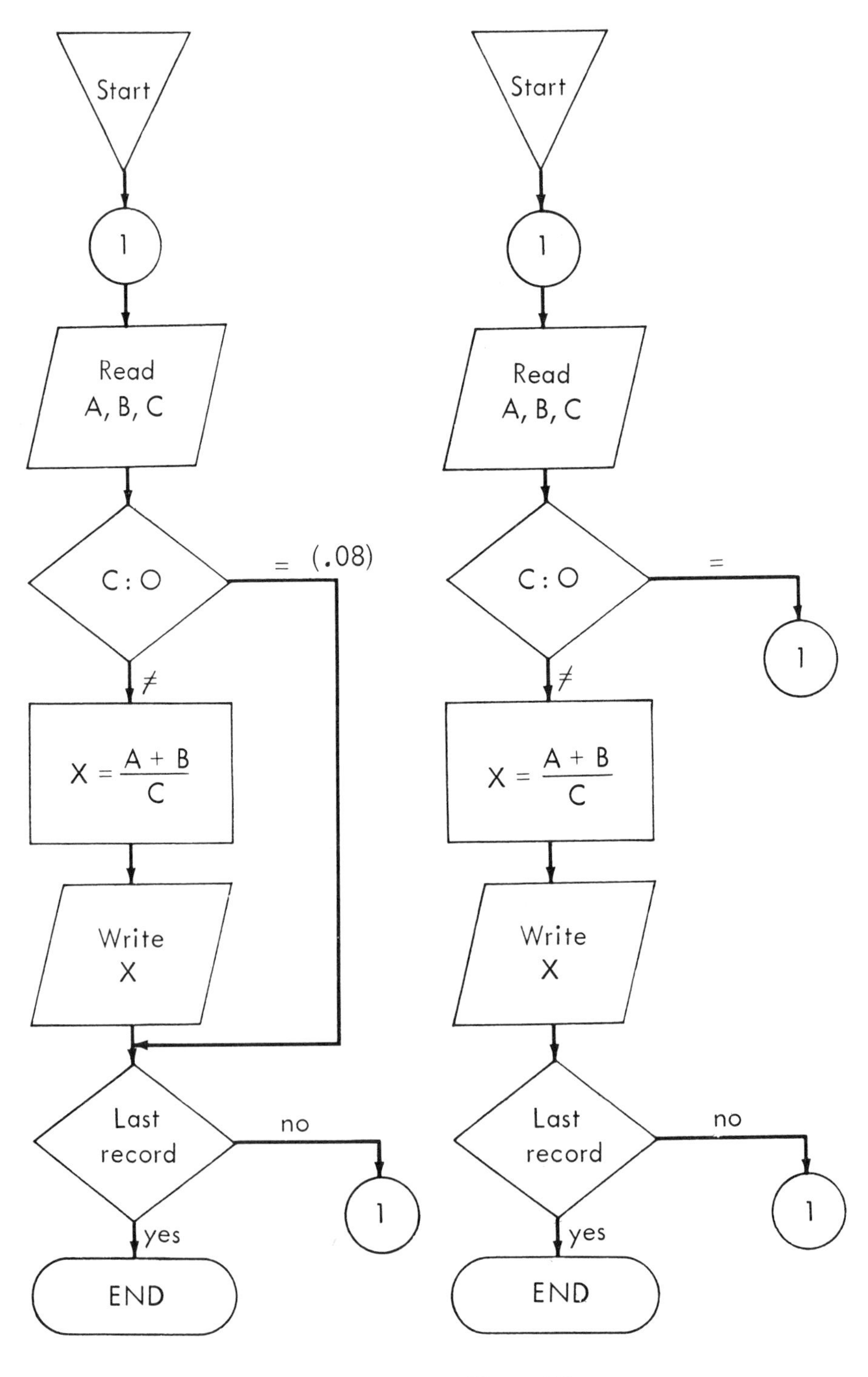

Figure 6-1

Figure 6-2

Figure 6–2 shows the same problem flowcharted incorrectly. In this case, the test for C = 0 branches back to read the next record, which will create an error condition if that record happens to be the last record. Caution should be used in the placement of a last record test. Notice that in Figure 6–1 the last record test is performed, regardless of the outcome of the other decision. When this "last record" procedure is used, the last record of the file must contain a special code to identify it as the last record.

Many problems will not have the last input record specially coded, but will state the total number of records to be processed. When this is the case, the programmer must count each record as it is processed and must terminate the processing after the final record has been processed. For example, a program has 50 input cards, each containing a single value. The program must select the largest value of the 50 input. The most efficient way to solve this problem is to successively compare two values, each time discarding the smaller value. The solution of this problem involving a counting operation is illustrated in Figure 6–3. The first step is to set the *counter* (C) to an initial value. In this case, the items to be counted are the number of cards read, so the counter is initially set to 1, and then the first card is read. The value from the first card (A) is assigned to the variable BIG as the largest value encountered so far. Then the loop is begun. The value B is read from the next card, and compared with BIG. If BIG is the larger, processing continues. If B is the larger, then BIG assumes the value of B (if they are equal, either path may be followed). At this point, BIG will contain the larger of the values from the first two cards. Regardless of the path followed from the decision, the value of C is incremented by 1 because another card has been read. Each time a card is processed and counted C is compared with 50 to determine if 50 cards have been read. If C is greater than or equal to 50, the value of BIG is output and processing is terminated. If C is less than 50, the flow loops back to the input step to read the next card, and the process repeats. In this manner, when C reaches the value of 50, BIG will contain the largest value of the input file.

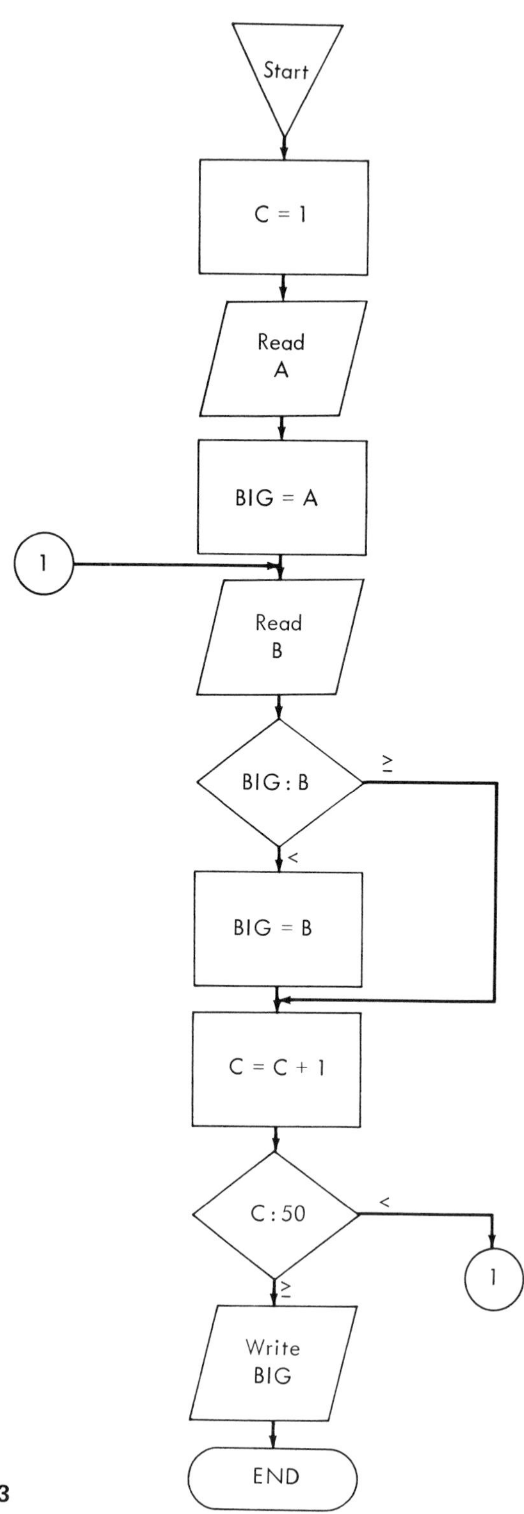

Figure 6-3

A note on computer technique – in the comparison of C with 50, C should never reach a value greater than 50. Nevertheless, the possibility should be taken into consideration on the basis that anything can possibly happen eventually will.

A counting operation consists of three distinct components: initialization, incrementation, and exit. In Figure 6–3, C = 1 is the *initialization* step, C = C + 1 is the *incrementation* step, and the comparison of C to 50 is the *exit* step. The programmer should be certain that each counting operation contains all three of these steps in the proper sequence. The counting process occurs so frequently in flowcharts that many programmers prefer to show all three steps within a single flowcharting symbol. Such a symbol is called an iteration box.[1]

Great care must be used to insure that the exit step directs the flow to the proper loop point. Consider the flowchart shown in Figure 6–4. This flowchart contains the three components but constitutes an infinite loop. Because of a return to an incorrect point, this flowchart instructs the computer to reset C to 1 on every loop through the program. Therefore, when the exit step is executed, C will always be equal to 2, thus causing the process to continue endlessly. The important point is that the initialization step must *never* be included in the steps to be repeated.

Many flowcharts involving a counting operation will also require an accumulation process which is similar to the counting procedure. Figure 6–5 illustrates the process of finding the sum of the first 20 integers. The variable SUM is an *accumulator* Notice that the variable SUM must be initialized and incremented.

In this process, the counter, C serves a dual purpose. In addition to its being used to terminate the processing, it also is used in the accumulation process. The counter is initially set to 1, since that is the first value to be processed, and SUM is set to 0. Then SUM assumes the value of the sum of SUM and C.

[1]For a discussion see, Philip M. Sherman, *Techniques in Computer Programming*. (Englewood Cliffs, N. J.: Prentice-Hall, Inc., 1970), pp 30–46.

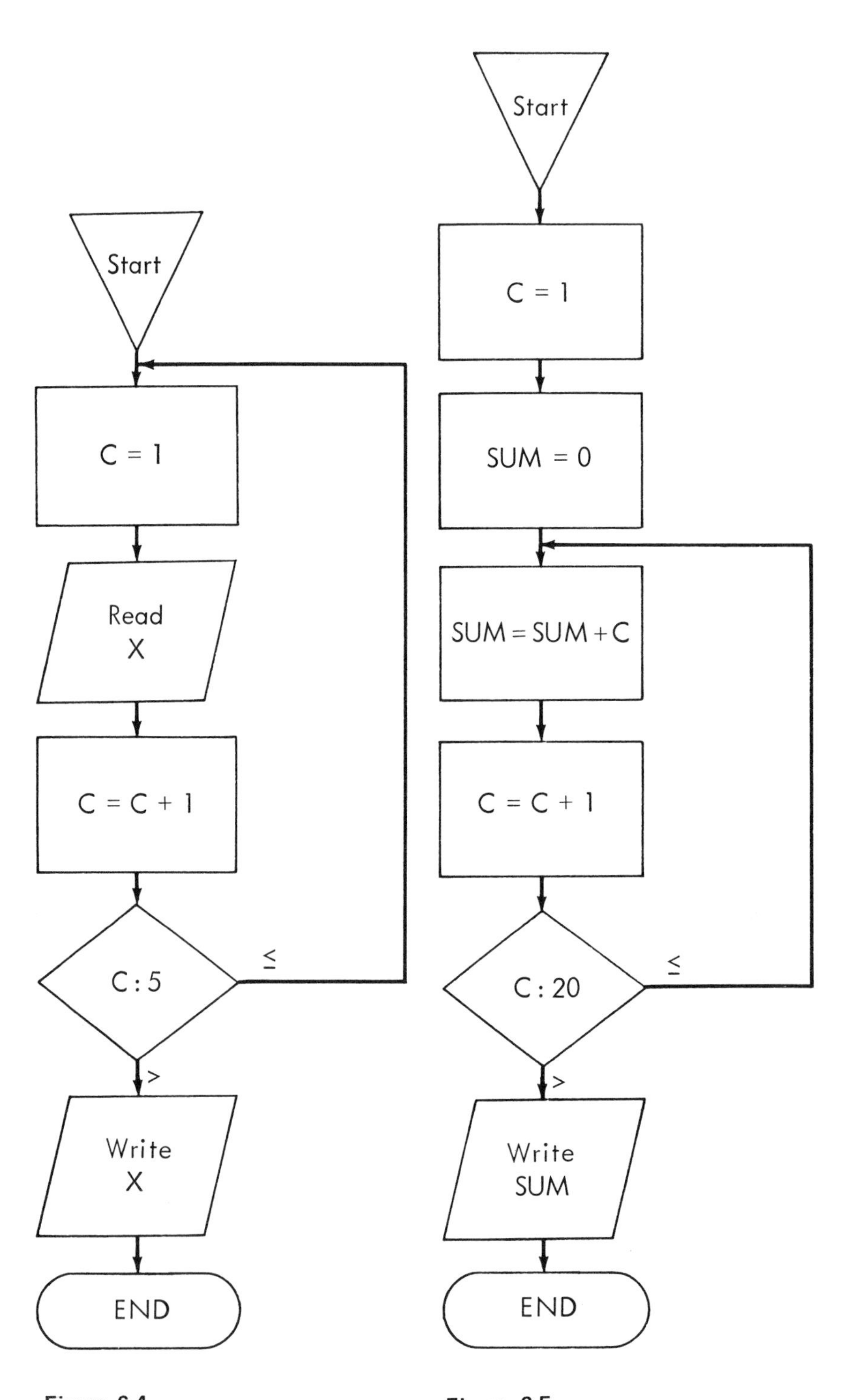

Figure 6-4

Figure 6-5

Table 6–1

	C	SUM
1st pass	1	0, then 1
2nd pass	2	1 + 2 = 3
3rd pass	3	3 + 3 = 6
4th pass	4	6 + 4 = 10
5th pass	5	10 + 5 = 15
6th pass	6	15 + 6 = 21
7th pass	7	21 + 7 = 28
8th pass	8	28 + 8 = 36
9th pass	9	36 + 9 = 45
10th pass	10	45 + 10 = 55
11th pass	11	55 + 11 = 66
12th pass	12	66 + 12 = 78
13th pass	13	78 + 13 = 91
14th pass	14	91 + 14 = 105
15th pass	15	105 + 15 = 120
16th pass	16	120 + 16 = 136
17th pass	17	136 + 17 = 153
18th pass	18	153 + 18 = 171
19th pass	19	171 + 19 = 190
20th pass	20	190 + 20 = 210

The first time through, SUM equals 0 and C equals 1, so SUM becomes 0 + 1 = 1. Then the value of C is incremented by 1, so C becomes 2. Then C is compared with 20. As C is less than 20, the process loops to the accumulation equation which causes SUM to become 1 + 2 = 3. Following the entire process through, C and SUM assume values as shown in Table 6–1.

The final value of C may be any desired value. At any point the processing is terminated, SUM will contain the sum at all integers up to that point.

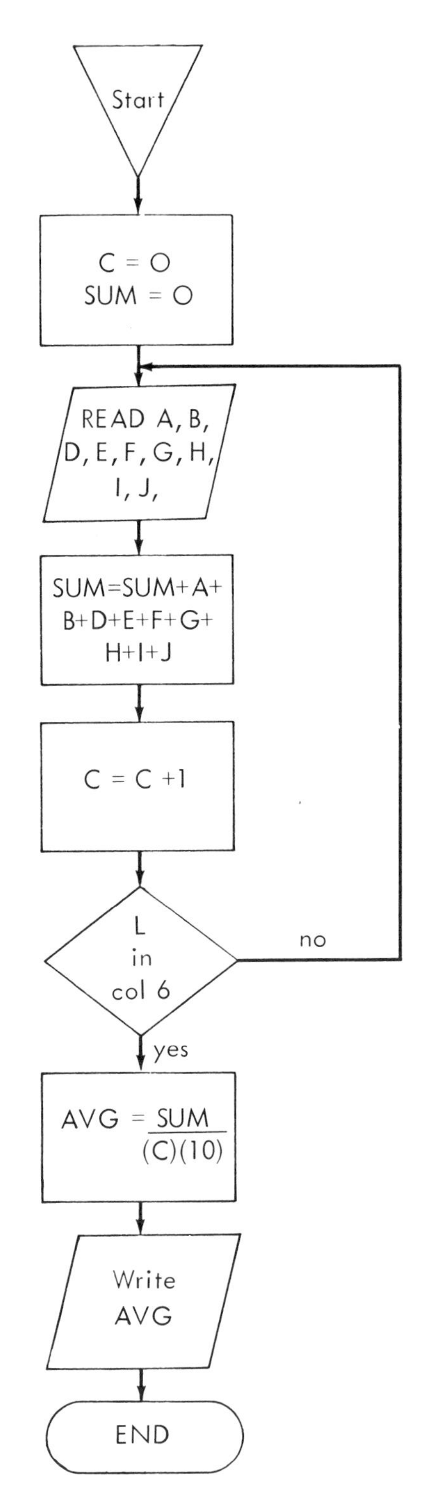

Start
C = O
SUM = O
READ A, B,
D, E, F, G, H,
I, J,
SUM=SUM+A+
B+D+E+F+G+
H+I+J
C = C +1
L
in
col 6
no
yes
AVG = SUM / (C)(10)
Write
AVG
END

Figure 6-6

Another type of problem will not state the number of input records but will require the programmer to have that information available for computation. Such problems require the last record to be specially coded. The programmer not only must count each record as it is read but also must test each record to determine if it is the final one.

A programmer is required to write a program with a set of input cards, each of which contains 10 numbers. He must compute the average of all the input values. He does not know how many cards there are, only that the last card contains an L in column 6. The solution to this problem is illustrated in Figure 6–6.

In this problem, the counter, C is initially set to zero and incremented as each card is read. As each card is processed, it is checked for an L in column 6. When the last card is encountered, the average of the accumulated value SUM must be computed, requiring the number of values read. Since C contains the number of cards processed, and there were 10 numbers per card, SUM is divided by C times 10 to obtain the desired average.

A final problem will illustrate many of the concepts of looping, counting, and accumulation plus a new idea. A bank arranges input cards as follows:

First card: This card contains the account number, name, and present balance.

Next card(s): Each card contains the account number, a "0" in column 6, and three deposit amounts.

Next card(s): Each card contains the account number, a "1" in column 6, and three withdrawal amounts.

Last card: The last card contains the account number, a "2" in column 6, and a service charge amount.

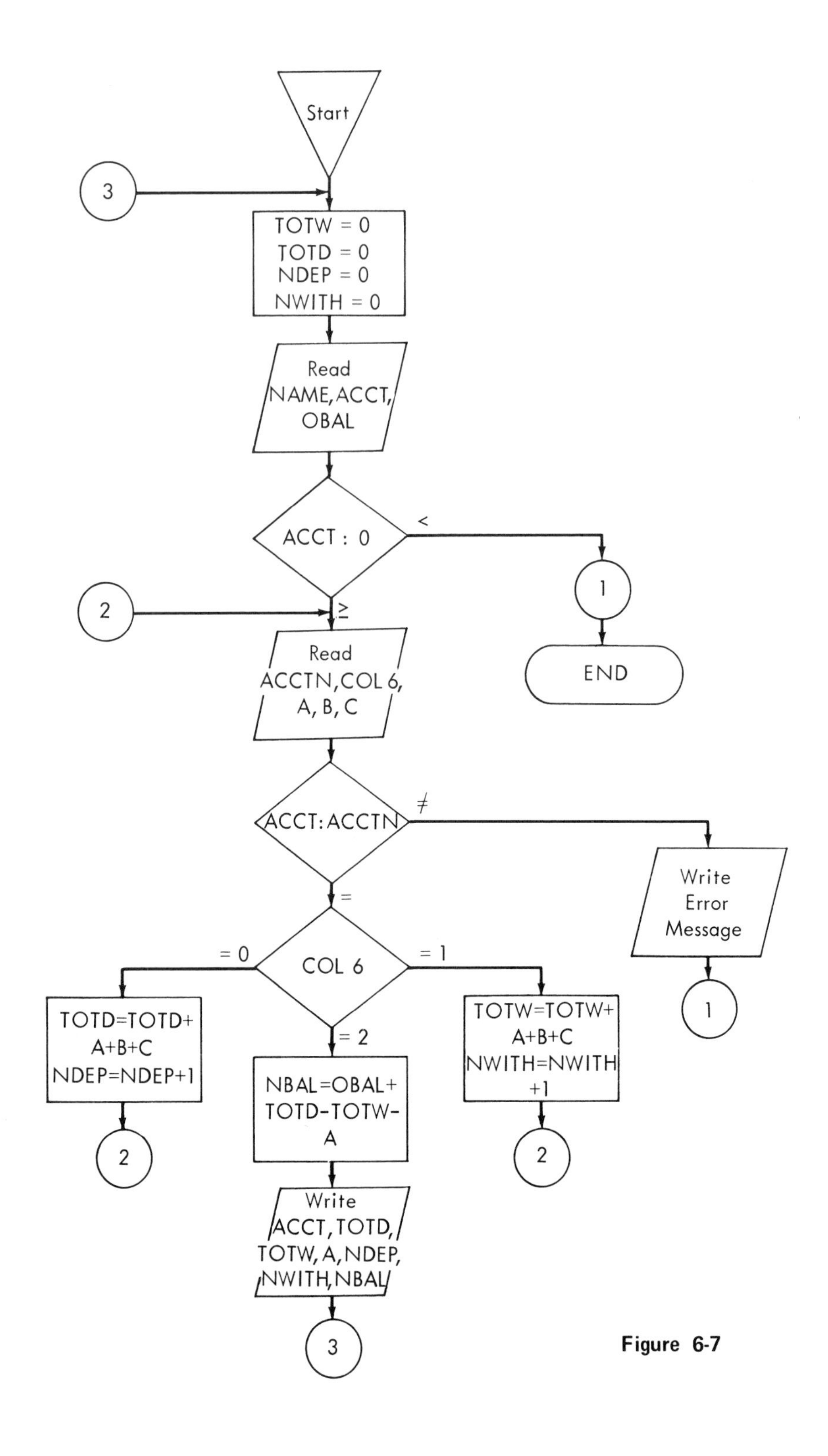

Start
3
TOTW = 0
TOTD = 0
NDEP = 0
NWITH = 0
Read
NAME, ACCT,
OBAL
ACCT : 0
<
1
END
2
≥
Read
ACCTN, COL 6,
A, B, C
ACCT: ACCTN
≠
Write
Error
Message
1
=
= 0
COL 6
= 1
TOTD=TOTD+
A+B+C
NDEP=NDEP+1
2
= 2
NBAL=OBAL+
TOTD-TOTW-
A
Write
ACCT, TOTD,
TOTW, A, NDEP,
NWITH, NBAL
3
TOTW=TOTW+
A+B+C
NWITH=NWITH
+1
2

Figure 6-7

The programmer does not know how many (if any) deposit and/or withdrawal cards are in each set. Also, he does not know in advance if the next card to be read is a deposit or a withdrawal card. The program will be terminated when a negative account number is encountered.

The programmer is required to have the computer read each set of inputs, compute the new balance, and produce a printed report showing the account number, total of deposits, total of withdrawals, service charge, number of deposits, number of withdrawals, and the new balance.

Figure 6–7 illustrates the solution to this problem. The first step is to initialize all quantities which must be accumulated; then, read the first card of a set of inputs. At this point, the account number is compared with zero. If the number is negative, processing is terminated. If the number is zero or positive, the next card is read, calling the account number ACCTN. To check for input errors, the account number from the first card (ACCT) is compared with ACCTN. If they are not equal the input is out of sequence or mispunched, and the program will halt with an appropriate error message being printed. If the account numbers are equal, then the code from column 6 is inspected to determine what type of card was read. If the code is 0 or 1, the amounts are added to the appropriate accumulator, and the process loops in order to read the next card in the input set. When the code number 2 is encountered, it indicates the last card of that set; so, the new balance is calculated, the output produced, and the process loops to the first card of the next input set. Processing will continue in this manner until either an error occurs or a negative account number is encountered. Notice that when one complete set of inputs is processed, the flowchart directs control to the initialization step, where the variables to be accumulated are reset to zero. This procedure will prevent carryover of the previous values into the next set of calculations.

EXERCISES

Draw flowcharts for the following problems:

1. Read a file of 50 values and count the number of negative values in the file.

2. Read a positive integer and compute its factorial. The factorial of $n, (n!)$ is defined as: $N! = 1 \cdot 2 \cdot 3 \cdot 4 \ldots (N-1) \cdot N$.

3. Determine how many one dollar bills, half dollars, quarters, dimes, nickels, and pennies to return in change for a five dollar bill after having read the amount of purchase.

4. An insurance policy increases in value at the rate of 8 percent per year for the first 10 years and at 10 percent per year thereafter. If the initial value of the policy is $1000, determine the value after N years.

5. Given that $f(x) = x^3 + 6x^2 - 3x + 4$, determine the value of $f(x)$ for values of x between 1 and 20 in increments of 0.5.

6. A college charges a tuition of $5 per hour to a maximum of $50. If a student enrolls for a laboratory course, he must pay a laboratory fee of $7.50. The input consists of a card for each course for which a student enrolls. The last card of each student's set is specially coded, as is the last card of the entire file. Compute the tuition due from each student and the total tuition due from all students.

7. The Acme Finance Company charges an interest rate of 1.25 percent per day on the unpaid balance of a loan. A payment of $1.00 per day is made. Show as output

how many days it would take to completely pay off a loan of x dollars.

8. A game is played with a jar containing 5 red balls, 5 black balls, and 1 white ball. Scoring is as follows:

   For first ball drawn:

   > Score 1 point if ball is black.
   >
   > Score 2 points if ball is red.

   For each succeeding ball drawn:

   > Score 1 point if ball is the same color as the preceding ball.
   >
   > Score 3 points if ball is not the same color as the preceding ball.

   When the white ball is drawn, the play ends, and the player drawing the white ball scores 3 points. Players A and B alternate turns drawing a ball. Draw a flowchart which will accumulate each player's score and produce the scores as output when the game ends.

7

Subscripted Variables

Many of the problems discussed in previous chapters have involved multiple sets of input data. In every case, the solution has been accomplished with only one set of data in storage at a time. In the problem of finding the largest of 50 values (Fig. 6–3), the values are read individually so that only one is available at any time. In some cases, this one-at-a-time approach will not suffice. For example, suppose that instead of 50 input cards, each with one number, there was only one card containing all 50 values. Because a card can only be read once all 50 values would have to be input at the same time. One method of resolving this situation is to specify 50 different variable names. This method is not desirable because of the tedious process of writing all the names; and, if each variable has a different name, the process could not be placed in a loop. The best method is to use a subscripted variable name. As in algebra, variables are subscripted to show different values at different times. The symbol $\mathbf{X}_t$ might represent the value of **X** at time t. Because no way exists to keypunch a subscript, computer nomenclature is a little different from algebraic nomenclature. Computer subscripts are usually written in parenthesis. Thus, the algebraic x_t

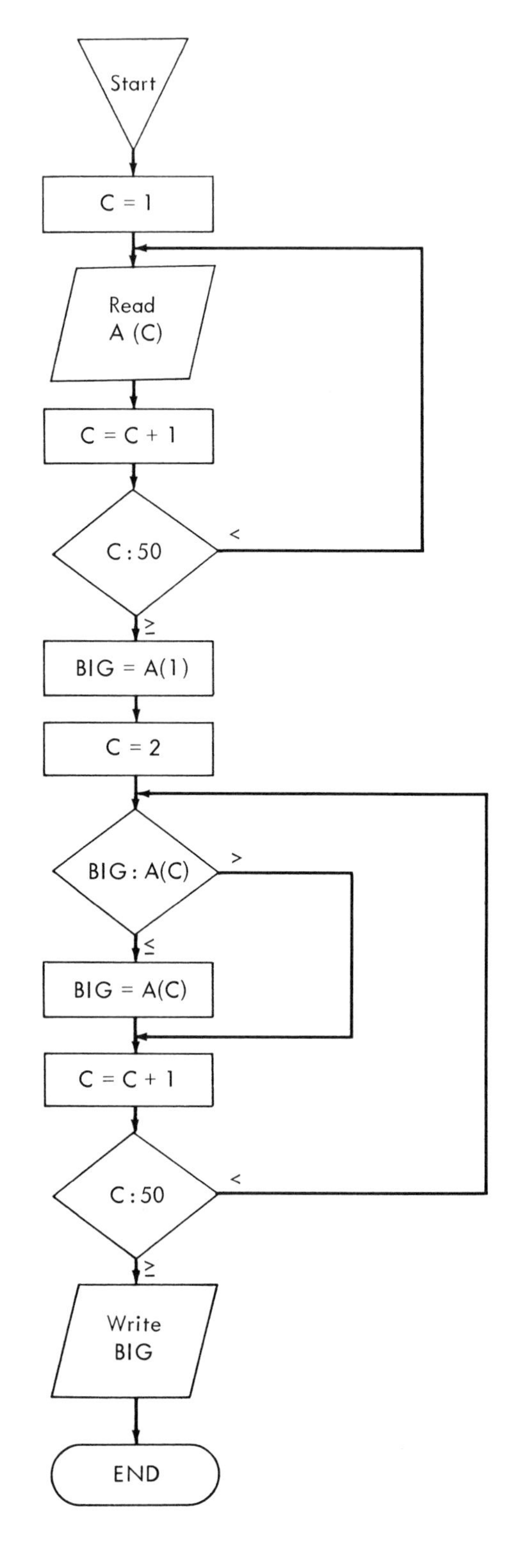
Start
C = 1
Read
A (C)
C = C + 1
C : 50
<
≥
BIG = A(1)
C = 2
BIG : A(C)
>
≤
BIG = A(C)
C = C + 1
C : 50
<
≥
Write
BIG
END

Figure 7-1

is equivalent to the computer's **X(T)**. In either notation, a subscript is simply a method of using the same name to denote multiple values. Subscripted variable names are sometimes called *array names*. Each item of data within the array is called an *element* of the array.

The use of an array for the solution of the largest value problem is illustrated in Figure 7–1. This solution uses an array named **A** to contain all 50 values. The counter is used as the subscript so that when **C** equals 1, the first value **A**(1) will be processed. When **C** equals 2, the second value **A**(2) will be processed and so forth. When the process of reading values into an array is flowcharted (Fig. 7–1), someone who is unfamiliar with the process might think that the values of **A** were being read from separate cards. To avoid this confusion and for ease of flowcharting, the process of reading values into an array is often shown as in Figure 7–2. The input block of this figure shows not only the array name, but also the range of values of the subscript.

One of the most difficult problems a beginning programmer encounters is the sorting of an array into ascending or descending order. One method of solution is to compare the first element with each additional element and to reverse their positions within the array when the next element is less than (or greater than) the first. This process will result in the smallest (or largest) element of the array occupying the first position. Then, the process is repeated, comparing the second element with all the remaining elements. In this manner, the smallest (or largest) element of the array will eventually occupy the first position, the second smallest (or largest) will occupy the second position, and so forth. This process is called an *interchange sort*. The interchange sort algorithm requires a loop within a loop, a process which occurs in many problems. The algorithm is flowcharted in Figure 7–3. After the array **A** is input, two separate counters are initialized, **C** and **K**. Then, **A(C)** is compared with **A(K)**. If **A(K)** is less than **A(C)**, their positions within the array are reversed. Then, K is incremented and tested. (**C** remains constant.) This process is repeated for all values of **K** up to 20. When **K** reaches 20, **C** is incremented, **K** set to 1 again, and the entire process is repeated.

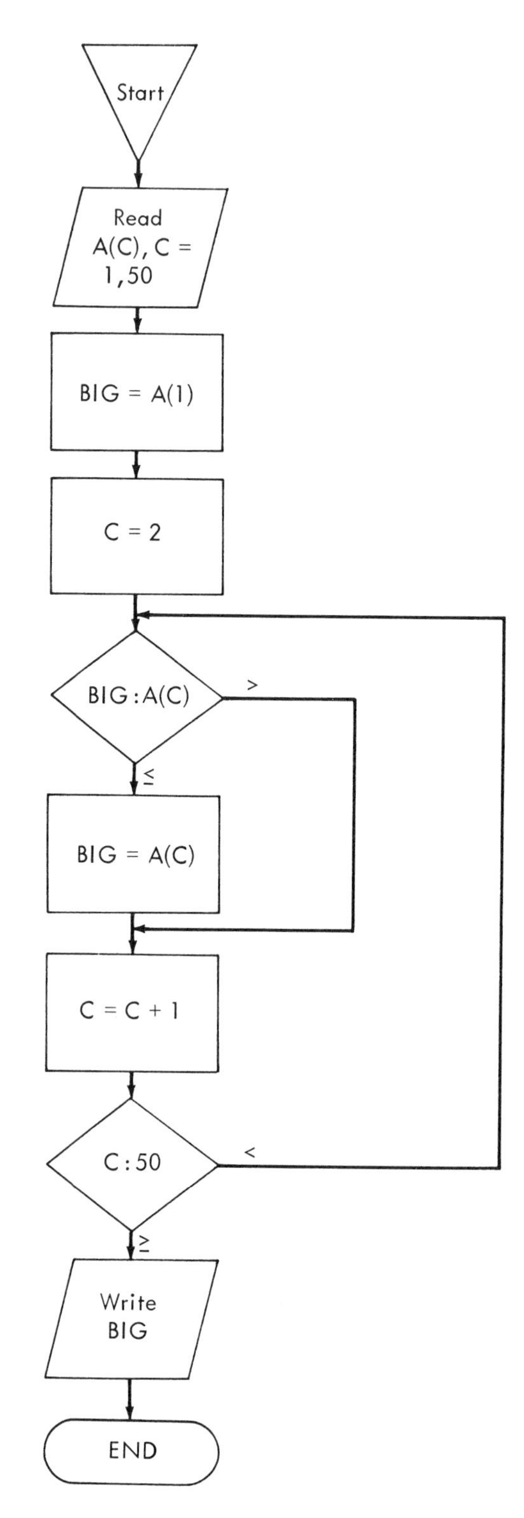

Figure 7-2

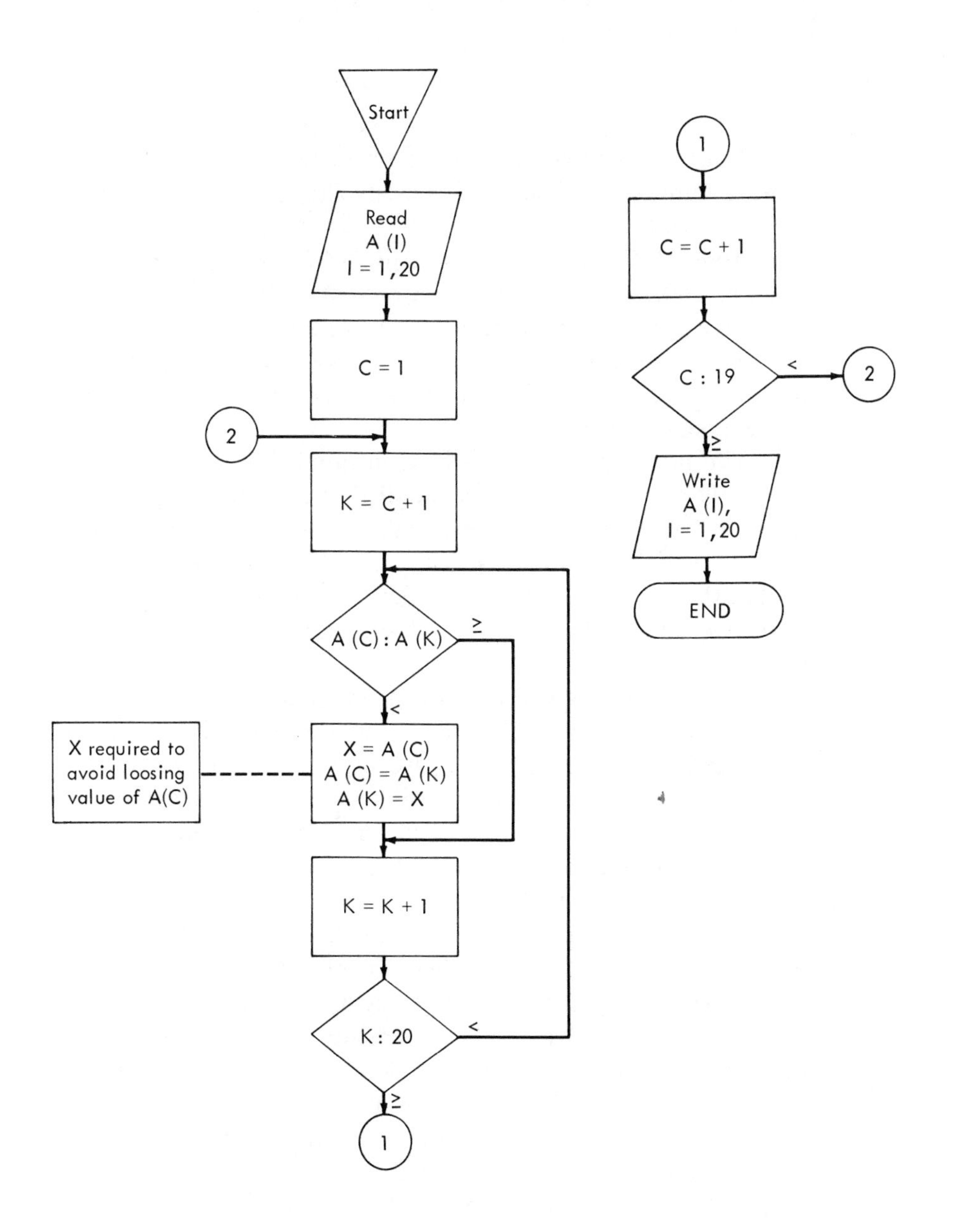
Start
Read
A (I)
I = 1,20
C = 1
2
K = C + 1
A (C) : A (K)
≥
<
X required to
avoid loosing
value of A(C)
X = A (C)
A (C) = A (K)
A (K) = X
K = K + 1
K : 20
<
≥
1
1
C = C + 1
C : 19
<
2
≥
Write
A (I),
I = 1,20
END

Figure 7-3

Notice that in the process of reversing the elements' positions, a temporary variable, **X**, is used. This is necessary because the statement **A(C) = A(K)** will erase the current value of **A(C)**. Therefore, **A(C)** is held temporarily by **X**, **A(C)** assumes the value of **A(K)**, and then **A(K)** assumes the value of **X**, which is what **A(C)** was before the reversing process began.

To insure understanding of this important loop-within-loop process, calculate the number of times the statement **K = K + 1** is executed in Figure 7–3.

The programmer should keep in mind that a card may be read only once. If the data is supplied as one number per card, then each number may be read individually. However, if there is more than one number per card, all of the numbers must be read at the same time. The best procedure is to read the numbers into an array and then to process them individually, as was done in Figure 7–2.

Many problems will require the programmer to work with more than one array. A typical example requires that the computer read 20 values from one card into an array named **A** and 20 values from a second card into array **B**. The processing will create an array **C**, each element of which is the average of the corresponding elements of **A** and **B**. The flowchart for this problem is shown in Figure 7–4.

In this flowchart, the elements of **A** are input first; then, the elements of **B** are input. Then, the elements of array **C** may be calculated one at a time by averaging the corresponding elements of **A** and **B**. When all 20 elements have been calculated, the elements of **C** are output and the processing terminates.

A problem requires a programmer to read a card containing 80 one-digit integers, determine the largest integer on the card, determine how many times it appears, and determine what columns of the card contain that integer. One method of solution, illustrated in Figure 7–5, is to read the integers into an array called NUM and then to search the array initially for nines. A variable called FLAG is used to count all the nines encountered. When all 80 values have been processed, FLAG is compared with zero. If FLAG is still equal to zero, COMP is reduced by 1, and the search process is repeated. When FLAG

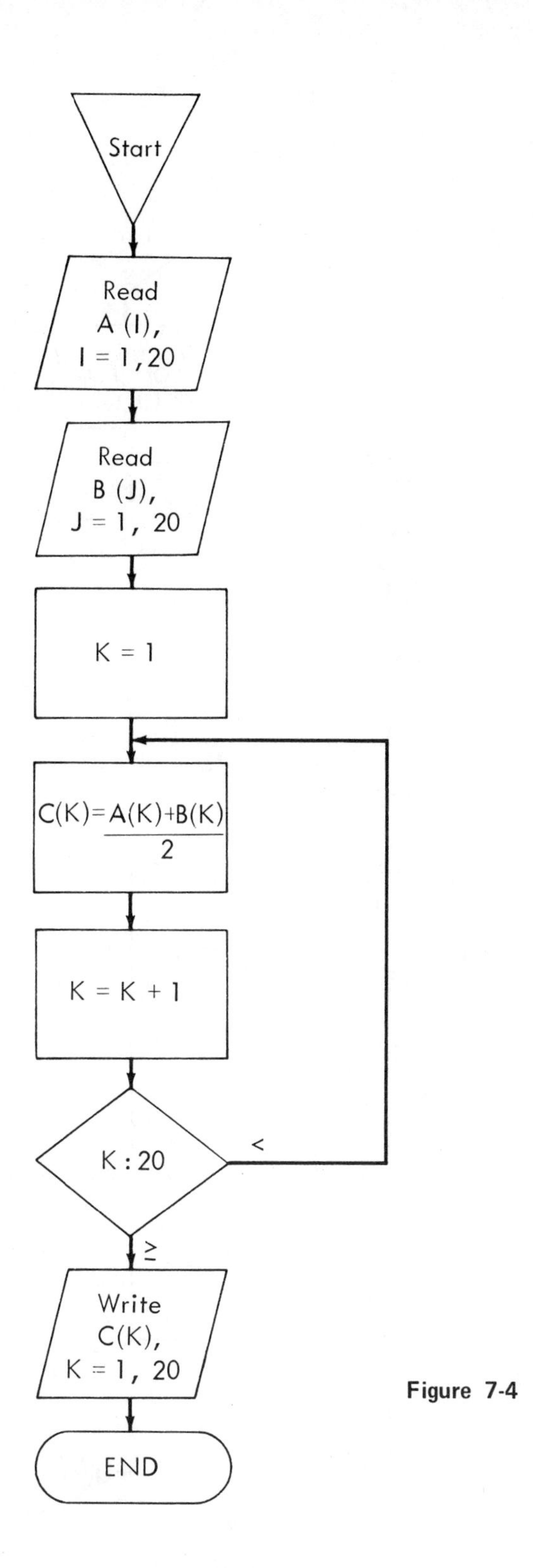
Start
Read
A (I),
I = 1, 20
Read
B (J),
J = 1, 20
K = 1
C(K)= A(K)+B(K) / 2
K = K + 1
K : 20
<
≥
Write
C(K),
K = 1, 20
END

Figure 7-4

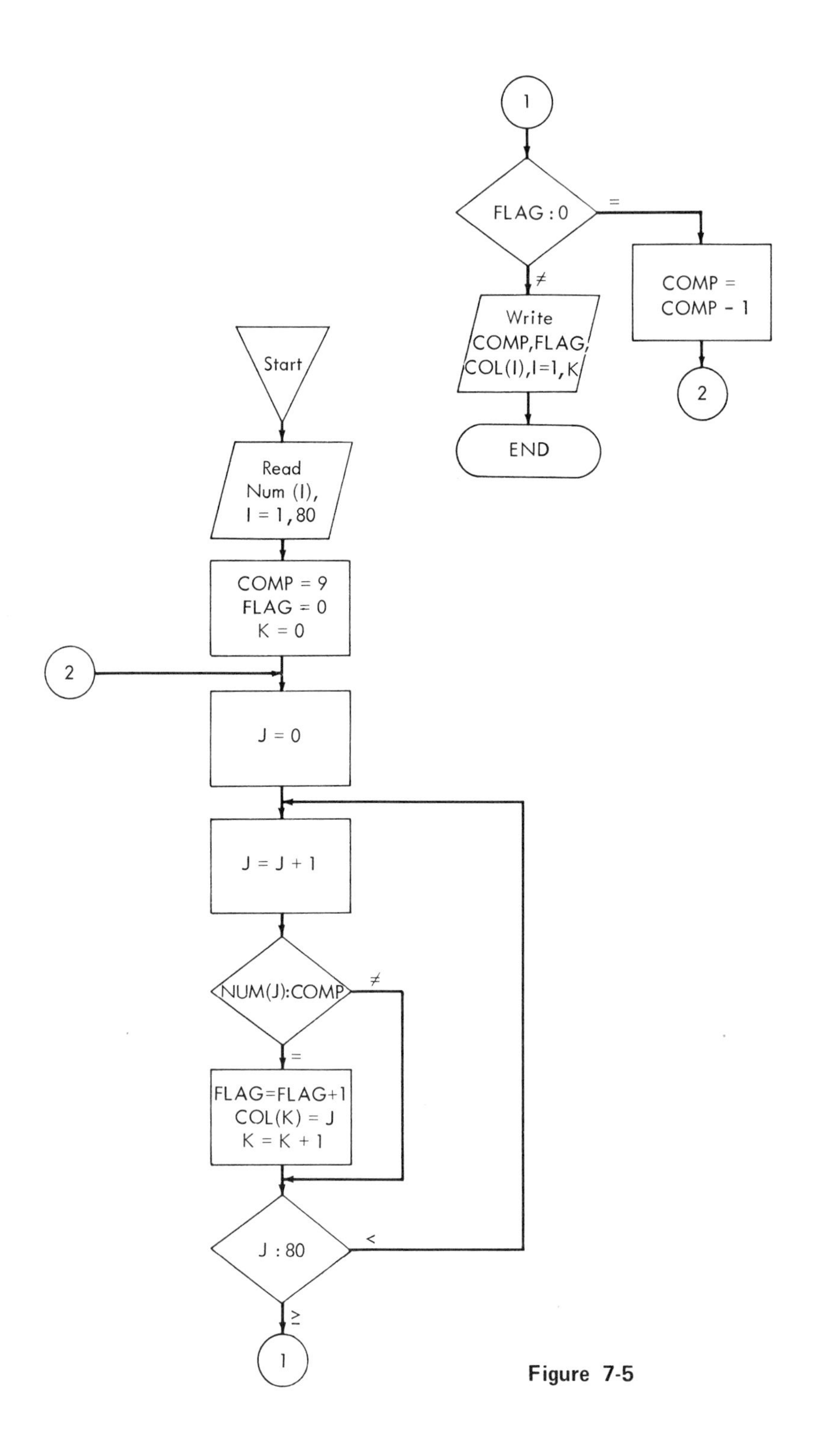
1
FLAG : 0
=
≠
COMP =
COMP - 1
Write
COMP,FLAG,
COL(I),I=1,K
2
END
Start
Read
Num (I),
I = 1,80
COMP = 9
FLAG = 0
K = 0
2
J = 0
J = J + 1
NUM(J):COMP
≠
=
FLAG=FLAG+1
COL(K) = J
K = K + 1
J : 80
<
≥
1

Figure 7-5

is found to be non-zero, this indicates that a larger value has been found. COMP will contain the largest value; FLAG will indicate the number of times COMP was contained in the card; COL (1) will indicate the first column containing COMP; COL (2) will indicate the next column containing COMP, and so forth.

EXERCISES

Draw the flowcharts for the following problems:

1. A card contains 10 item stock numbers. A second card contains the corresponding amounts of stock on hand. A third card contains the corresponding amounts of stock issued. Compute the current stock on hand for each item.

2. A card contains the radii of five different spheres. Use the formula $V = 4/3\pi r^2$ to compute the volume of each sphere. Also, the average of the five volumes must be shown as output.

3. Solve the problem illustrated in Figure 6–7 by using an array to contain the withdrawals, deposits, and service charges as they are read from the card.

4. Given two arrays, **A** and **B**, each consisting of 40 elements, interchange the elements of the two arrays and reverse their order within the array. For example, place **A**(1) into **B**(40) and **B**(40) into **A**(1); **A**(2) into **B**(39) and **B**(39) into **A**(2) and so forth. Remember that once an element is replaced, it is lost; so, the process **A**(1) = **B**(40), **B**(40) = **A**(1) will simply make both **A**(1) and **B**(40) equal to whatever **B**(40) was initially.

5. Read a list of 100 values into an array; then, read a single value **A**. The computer is required to select the value from the array that exceeds **A** by the greatest amount.

6. Develop another algorithm for sorting an array into ascending order.

7. A class has 50 students. The instructor plans to perform a statistical analysis of a set of test scores to determine the mode, median, and mean of the scores. When the scores are arranged in descending order, the mode is the score which appears most often, the median is the "middle" score, and the mean is the average score. Read the scores from a card, sort them into descending order, and compute the three output values.

8. An array is to be generated in the following manner. The first element is to be read from a card, and each successive element will be equal to ½ the preceding element. A process must be defined which will read the first element and will compute the successive elements of the array until an element less than 0.001 is computed. Show as output the number of elements that the array contains; also, show the elements.

9. The trigonometric sine of an angle x may be computed by the formula:

$$\text{sine } x = x - \frac{x^3}{3} + \frac{x^5}{5} - \frac{x^7}{7} + \cdots$$

Develop an algorithm which will compute the sine of an input angle to within 0.0001 of the actual value.

8

Is it Right?

The purpose of drawing a flowchart is to make the coding of the problem easier. The program code should follow the flowchart step by step. When this procedure is followed, the program code should reflect exactly the same procedures as those of the flowchart. Obviously, if the flowchart is incorrect, the program will be coded incorrectly. Therefore, the programmer should be sure his flowchart is drawn properly before coding. Determining whether the flowchart is correct or not may prove to be a difficult task.

As stated previously, there is no one right flowchart for a given problem. If a flowchart leads to the desired result, then it is correct. However, this is not meant to imply that there are no *wrong* flowcharts. It is possible to draw a correct flowchart which solves the wrong problem. After the flowchart is drawn, the programmer must objectively determine if the procedures used do indeed produce that which was intended.

The programmer must test his procedures with a simple set of input data. A set of data should be chosen which will allow the programmer to anticipate the correct answer. Then, he may manipulate the data as instructed by the flowchart and

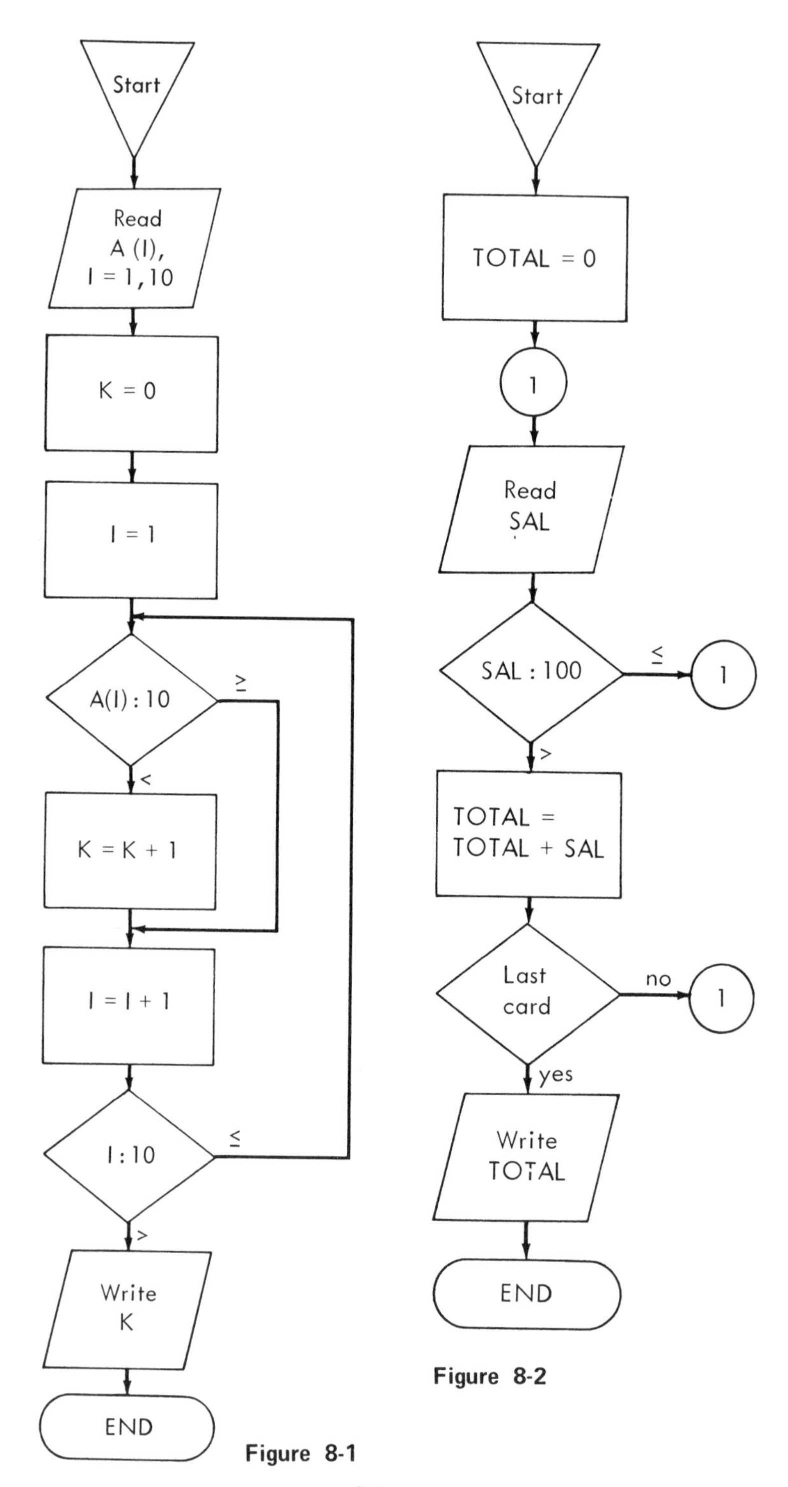

Figure 8-1

Figure 8-2

determine if it leads him to the correct result. This process involves keeping a table which shows the current value of each variable used in the flowchart. A flowchart of a problem to determine the number of values less than 10 in an input array of 10 values is shown in Figure 8–1. To test the flowchart, the programmer would make up a list of 10 values and count the number of values which are less than 10. Table 8–1 shows the process of verifying that the flowchart does count the values properly. The flowchart (Fig. 8–1) indicates that the final value of **K** is the answer. In Table 8–1, the final value of **K** is four, which is the correct result.

Table 8–1

Input	K	I
	0	1
A(1)=4	1	2
A(2)=3	2	3
A(3)=12	2	4
A(4)=24	2	5
A(5)=36	2	6
A(6)=19	2	7
A(7)=10	2	8
A(8)=8	3	9
A(9)=14	3	10
A(10)=6	4	11

The construction of a table verifying a flowchart involves the programmer in "playing computer," that is, he must do only what he is instructed to do by the flowchart. It is easy to fall into the trap of assuming something which is not in the flowchart.

Consider Figure 8–2. The object of this flowchart is to total all the salaries which are less than $100 and to print the total obtained when all the records have been read. A sample set of inputs and the verifying table are shown in Table 8–2(a). This set of data does not allow all the possible cases. One verification does not prove that the flowchart is totally correct. Consider the set of data and the table shown in Table 8–2(b).

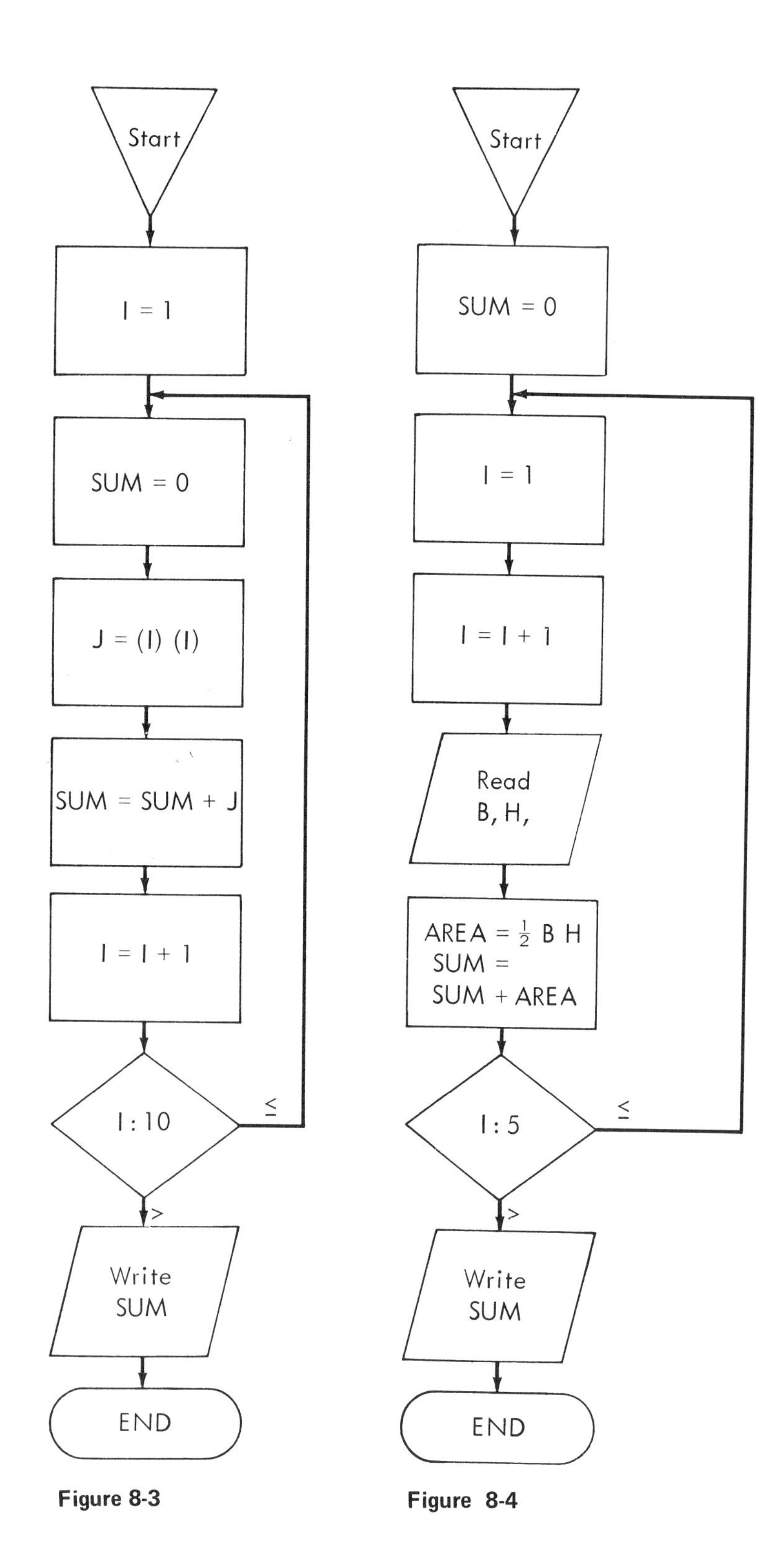

Figure 8-3

Figure 8-4

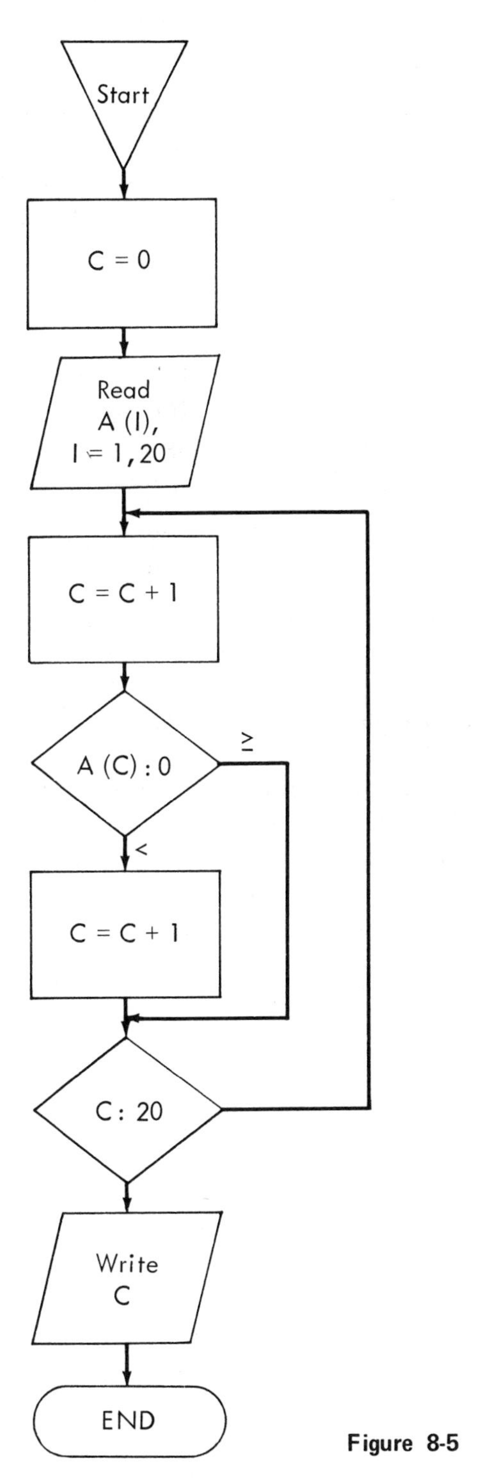

Figure 8-5

Table 8–2(a)

Input: 101,200,98,48,300
Output: Total = 601

SAL	Total
	0
101	101
200	301
98	301
48	301
300	601

Table 8–2(b)

Input: 101,200,98,300,48
Output: Total = 601

SAL	Total
	0
101	101
200	301
98	301
300	601
48	601
0	

The last value shown for **SAL** is zero because the flowchart instructs the computer to read another record if the last record contained a salary greater than or equal to $100. In this case, the physical arrangement of the data causes an error in the flowchart.

The flowchart of a problem designed to find the sum of the squares of the first 10 intergers is shown in Figure 8–3. Table 8–3 illustrates the result of the flowcharting error of including the initialization step in the loop. Table 8–3 clearly indicates that the final value of the variable **SUM** is not the desired 385.

Table 8–3

I	J	Sum
1	1	0
2		1
	4	0
3		4
	9	0
4		9
	16	0
5		16
	25	0
6		25
	36	0

I	J	Sum
7		36
	49	0
8		49
	64	0
9		64
	81	0
10		81
	100	0
11		100

EXERCISES

1. Figure 8–4 shows a flowchart of a procedure to calculate the area of five triangles and the sum of the areas. Determine if the flowchart is correct.

2. Figure 8–5 shows a flowchart of a procedure to count the number of negative elements in an array. Determine if the flowchart is correct.

3. Prepare a sample set of inputs and verify that Figure 6–6 is correct.

4. Prepare a sample set of inputs and verify that Figure 7–3 is correct.

5. A programmer has been employed by a basketball team to maintain statistics for each player for each game. For each game, a card is provided for each player containing his name, field goals attempted, field goals made, free throws attempted, free throws made, rebounds, and fouls. Output is to consist of all input data, field goal percentage, free-throw percentage, and points scored (field goal = 2 points, free throw = 1 point). The last card contains a dummy name and a negative value for field goals attempted. Flowchart this problem and verify that the flowchart is correct.

6. After the program for problem 5 above was written, the team decided to include season-to-date statistics by providing a tape containing season totals for each player. As each player's card is read, a matching tape record is read, game data computed, season-to-date data computed, and the tape updated with the new totals. The season-to-date output is to consist of field goal percentage, free throw percentage, field goals attempted, field goals made, free throws attempted, free

throws made, total rebounds, average rebounds per game, total fouls, average fouls per game, total points, and average points per game. Modify the flowchart to include this change.

7. Draw a flowchart for the following problem. An oil company classifies its credit card holders as to resident of the city, resident of the state, or out-of-state resident. A programmer must punch a code "01" in the city resident cards, "02" in the state resident cards, and "03" in the out-of-state resident cards. The number of cards of each type is unknown. All the cards of each type are grouped together in the input deck. The program is to read the input deck and punch the proper code into each card.

8. Draw a flowchart for the following problem. A college computing center receives a grade card for each student for each course in which he is enrolled containing his name, student number, course number, and the grade as A, B, C, D, or F. The course number is a three-digit integer; the tens digit indicates the credit hours of the course. Output is to consist of hours attempted, quality point earned, and the grade point average.

APPENDIX A

Number Systems

In dealing with computer applications, data must often be converted into number systems which the computer can manipulate more readily. These conversions involve primarily the binary (base 2) and the hexadecimal (base 16) systems. The purpose of this appendix is to provide a brief review of the various conversion techniques.

Binary-to-Decimal Conversion

A binary number may be converted to its decimal equivalent by multiplying each digit of the binary number by the appropriate power of 2 (Table A–1), starting with 2^0 at the right-most digit, using ascending powers, and proceeding from right to left.

Table A–1 - Powers of 2

n	2^n
0	1

1	2
2	4
3	8
4	16
5	32
6	64
7	128
8	256
9	512
10	1024
11	2048
12	4096
13	8192
14	16384
15	32768

Example

$$\begin{aligned} 10010_2 &= 1 \cdot 2^4 + 0 \cdot 2^3 + 0 \cdot 2^2 + 1 \cdot 2^1 + 0 \cdot 2^0 \\ &= 1 \cdot 16 + 0 \cdot 8 + 0 \cdot 4 + 1 \cdot 2 + 0 \cdot 1 \\ &= 16 + 0 + 0 + 2 + 0 \\ &= 18_{10} \end{aligned}$$

Example

$$\begin{aligned} 1010111_2 &= 1 \cdot 2^6 + 0 \cdot 2^5 + 1 \cdot 2^4 + 0 \cdot 2^3 + 1 \cdot 2^2 + 1 \cdot 2^1 + 1 \cdot 2^0 \\ &= 1 \cdot 64 + 0 \cdot 32 + 1 \cdot 16 + 0 \cdot 8 + 1 \cdot 4 + 1 \cdot 2 + 1 \cdot 1 \\ &= 64 + 0 + 16 + 0 + 4 + 2 + 1 \\ &= 87_{10} \end{aligned}$$

Decimal-to-Binary Conversion

A decimal number may be converted to its binary equivalent by reversing the binary-to-decimal conversion process. This is accomplished by writing the decimal value as a sum of numbers which are powers of 2 and placing 1's in the corresponding positions in a binary string.

Example

$$
\begin{aligned}
23_{10} &= 16 + 4 + 2 + 1 \\
&= 1 \cdot 16 + 0 \cdot 8 + 1 \cdot 4 + 1 \cdot 2 + 1 \cdot 1 \\
&= 1 \cdot 2^4 + 0 \cdot 2^3 + 1 \cdot 2^2 + 1 \cdot 2^1 + 1 \cdot 2^0 \\
&= 10111_2
\end{aligned}
$$

Example

$$
\begin{aligned}
17_{10} &= 16 + 1 \\
&= 1 \cdot 16 + 0 \cdot 8 + 0 \cdot 4 + 0 \cdot 2 + 1 \cdot 1 \\
&= 1 \cdot 2^4 + 0 \cdot 2^3 + 0 \cdot 2^2 + 0 \cdot 2^1 + 1 \cdot 2^0 \\
&= 10001_2
\end{aligned}
$$

Hexadecimal-to-Decimal Conversion

A hexadecimal number may be converted to its decimal equivalent by converting each digit to its decimal equivalent (Table A–2), multiplying each digit by the appropriate power of 16 (Table A–3), starting with 16^0 at the right most digit, using ascending powers, and proceeding from right to left.

Table A–2
Hexadecimal, Binary and Decimal Equivalents

Hexadecimal	Decimal	Binary
0	0	0000
1	1	0001
2	2	0010
3	3	0011
4	4	0100
5	5	0101
6	6	0110
7	7	0111
8	8	1000
9	9	1001
A	10	1010

B	11	1011
C	12	1100
D	13	1101
E	14	1110
F	15	1111

Table A–3 Powers of 16

n	16^n
0	1
1	16
2	256
3	4096
4	65536
5	1048576
6	16777216
7	268435456
8	4294967296
9	68719476736
10	1099511627776

Example

$$
\begin{aligned}
18F_{16} &= 1 \cdot 16^2 + 8 \cdot 16^1 + F \cdot 16^0 \\
&= 1 \cdot 256 + 8 \cdot 16 + 15 \cdot 1 \\
&= 256 + 128 + 15 \\
&= 399_{10}
\end{aligned}
$$

Example

$$
\begin{aligned}
2A4C_{16} &= 2 \cdot 16^3 + A \cdot 16^2 + 4 \cdot 16^1 + C \cdot 16^0 \\
&= 2 \cdot 4096 + 10 \cdot 256 + 4 \cdot 16 + 12 \cdot 1 \\
&= 8192 + 2560 + 64 + 12 \\
&= 10828_{10}
\end{aligned}
$$

Decimal-to-Hexadecimal Conversion

A decimal number may be converted to its hexadecimal equivalent by reversing the hexadecimal-to-decimal process. This is accomplished by writing the decimal value as a sum of numbers which are multiples of powers of 16 and placing the coefficients in the corresponding positions in a hexadecimal string.

Example

$$\begin{aligned} 163_{10} &= 160 + 3 \\ &= 10 \cdot 16 + 3 \cdot 1 \\ &= \mathrm{A} \cdot 16^1 + 3 \cdot 16^0 \\ &= \mathrm{A3}_{16} \end{aligned}$$

Example

$$\begin{aligned} 1258_{16} &= 1024 + 224 + 10 \\ &= 4 \cdot 256 + 14 \cdot 16 + 10 \cdot 1 \\ &= 4 \cdot 16^2 + \mathrm{E} \cdot 16^1 + \mathrm{A} \cdot 16^0 \\ &= \mathrm{4EA}_{16} \end{aligned}$$

Hexadecimal-to-Binary Conversion

A single hexadecimal digit may be written as a sequence of four binary digits (Table A–2). To convert a hexadecimal value to its binary equivalent, replace each hexadecimal digit with its four-digit binary equivalent.

Example

$$\mathrm{26AF}_{16} = \frac{0010}{2}\ \frac{0110}{6}\ \frac{1010}{\mathrm{A}}\ \frac{1111}{\mathrm{F}} = 10011010101111_2$$

Example

$$\mathrm{BC08}_{16} = \frac{1011}{\mathrm{B}}\ \frac{1100}{\mathrm{C}}\ \frac{0000}{0}\ \frac{1000}{8} = 1011110000001000_2$$

Binary-to-Hexadecimal Conversion

To convert a binary number to its hexadecimal equivalent starting at the right, arrange the binary value into groups of four digits, and replace each group with its hexadecimal equivalent (Table A–2).

Example

$$100101000000_2 = \frac{0100}{4}\ \frac{1010}{A}\ \frac{0000}{0} = 4A0_{16}$$

Example

$$1001011111110110_2 = \frac{1001}{9}\ \frac{0111}{7}\ \frac{1111}{F}\ \frac{0110}{6} = 97F6_{16}$$

APPENDIX B

Binary Logic

The basis of most computer circuitry is binary logic which consists of the three operations AND (∧), OR (∨), and NOT (¬) and various combinations of these operations. In the following discussions, A and B represent any proposition.

The NOT, Operation

The premise of the NOT operation is that the "NOT of A" (¬A) is true when A is false, and false when A is true.

Table B—1 The NOT Operation

A	¬A
T	F
F	T

The AND Operation

The premise of the AND operation is that the combination "A AND B" ($A \wedge B$) is true when *both* A and B are true, and false if either A or B, or both, are false.

Table B–2 The AND Operation

A	B	$A \wedge B$
T	T	T
F	T	F
T	F	F
F	F	F

The OR Operation

The premise of the OR operation is that the combination "A OR B" ($A \vee B$) is true when *either* A or B or both is true, and false if both A and B are false.

Table B–3 The OR Operation

A	B	$A \vee B$
T	T	T
F	T	T
T	F	T
F	F	F

Glossary

Accumulator - A variable designed to maintain a current value of a total in a looping process.

Algorithm - A description of a detailed procedure to solve a problem.

Annotation - The process of providing explanatory notes.

Array - A list of values occupying several storage locations.

Card - A heavy paper of standard size used to record data in the form of holes punched into the paper.

Card Reader - A device capable of converting holes in a card into computer electronic codes.

Code - (1) A set of meaningful symbols, (2) the actual program instructions written in a symbolic language, and (3) the process of writing instructions in a symbolic language.

Computer - A device capable of performing basic arithmetical computations on given information and retaining the results.

Counter - A variable used to determine the number of times an instruction is executed.

Data - Meaningful information.

Data Processing - The procedure of arranging given data into a desired result.

Decision - The computer process of comparing two values.

Decision Box - The symbol used in flowcharting to indicate that a comparison (decision) is to be made.

Disk (Magnetic) - A storage device similar to a phonograph record in that data may be recorded on it, and retrieved from it.

Documentation - A complete description of a computer program, consisting of word descriptions, and all three levels of flowcharts.

Element - Each item of data within an array.

File - A collection of related data input or output through one external device.

Flow - The term used to describe the sequence of execution of procedures in a given problem.

Flowchart - A graphic representation of the procedures leading to the solution of a given problem.

Initialize - To set a variable or counter to a starting value.

Input - (1) The process of supplying data to the computers storage, (2) the actual data to be supplied.

Instruction - A command to be executed by the computer.

Keyboard - A typewriter-like input device through which data may be entered into storage.

Keypunch - The process of placing data onto cards in the form of punched holes.

Location - The place reserved in storage for a particular item of data.

Loop - A set of instructions to be executed repeatedly. The loop must consist of the elements initialization, process, and exit.

Output - (1) The process of obtaining data from storage, (2) the actual data so obtained.

Printer - An output device capable of receiving data from a computer and printing the data on paper.

Program - An ordered set of instructions designed to provide the solution to a problem.

Read - To place data into storage.

Record - A set of data pertaining to a particular item.

Sort - To arrange data into some predefined order.

Storage - The part of the computer which retains information. The term is preferable to memory.

Storage location - Same as location.

Subscript - A number which denotes one of several items in a list.

Tape (Magnetic) - A strip of material sensitive to magnetic impulses on which data may be stored coded as impulses.

Tape (Paper) - A strip of heavy paper on which data may be stored in the form of punched holes.

Variable - A symbol which denotes a particular storage location.

Write - To obtain data from storage.

Bibliography

Arnold, Robert R., Harold C. Hill, and Aylmer V. Nichols, *Modern Data Processing.* New York, N. Y.: John Wiley & Sons, Inc., 1969.

Farina, Mario V., *Flowcharting..* Englewood Cliffs, N. J.: Prentice-Hall, Inc., 1970.

Saxon, James A. and Wesley W. Steyer, *Basic Principles of Data Processing.* Englewood Cliffs, N. J.: Prentice-Hall, Inc., 1970.

Sherman, Philip M., *Techniques in Computer Programming.* Englewood Cliffs, N. J.: Prentice-Hall, Inc., 1970.

Wheeler, Gershon J., and Donlan F. Jones, *Business Data Processing: An Introduction.* Reading, Mass.: Addison-Wesley Publishing Co., Inc., 1966.

Index